# VEGETABLES

Rodale's Home Gardening Library™

# VEGETABLES

edited by Anne M. Halpin

Rodale Press, Emmaus, Pennsylvania

Printed in the United States of America on recycled paper containing a high percentage of de-inked fiber (black and white pages only).

Book design by Marcia Lee Dobbs and Julie Golden
Illustrations by Kathi Ember and Jack Crane
Photography credits: J. Michael Kanouff: photo 10; Rodale Press Photography Department: photos 1, 2, 3, 4, 5, 6, 7, 8, 9

**Library of Congress Cataloging-in-Publication Data**

Vegetables.
(Rodale's home gardening library)
1. Vegetable gardening. 2. Organic gardening.
3. Vegetables. I. Halpin, Anne Moyer. II. Series.
SB324.3.V44 1988 635 87-23311
ISBN 0-87857-737-8 paperback

2 4 6 8 10 9 7 5 3 1 paperback

# Contents

# 1

# Getting Started

Long before the shovel slides into your garden soil and a single seed is sown, you should already have a clear vision of what your garden is going to look like, both in terms of its size and of what vegetables will be found there. A good place to start is to ask yourself the following questions:

How much space do you have? A country dweller with several acres obviously has more room to work with than a suburbanite who is limited to a 5-by-10-foot patch along the driveway. But whether you're dealing in feet or acres, both are viable gardening spaces. Even apartment and city dwellers can transform a variety of containers on a sunny roof or patio into a flourishing, productive garden.

How much space and time do you want to devote to the garden? You may have 1,000 square feet at your disposal, but that doesn't mean you should immediately plan to plant the entire area; that may be more of an investment in time than you are ready to make. There's no way around it—gardening does take time, if you want to do it well. There are ways to cut down on maintenance, but the truth of the matter is that every square foot of garden is going to require your attention in the course of preparing the soil, planting, mulching, watering, and so on throughout the season. If the time you can spend on a garden is limited, keep this in mind as you plan the dimensions. It's better to err on the small side the first year than to prepare a large area only to watch it become overgrown and unkempt as the season progresses. After successfully managing a small garden for one season, you can add on as your gardening experience grows.

What are your family's likes and dislikes? It's a waste of time, money, and garden space to plant a vegetable your family won't eat, unless you plan on marketing it. It's only common sense to concentrate your efforts on vegetables that will be enjoyed, and not to tie up garden space with crops that will be pushed to the side of the plate. If you're trying out a new vegetable, start sparingly. If it's a success, you can plant more next season; if it doesn't go over well, you haven't invested a large amount of productive space on a failure.

## Choosing the Site

Choosing the location for your vegetable garden is the first exciting and imaginative step on the way to actually making it come alive. It's not necessary to begin with virgin land. If there's an old garden plot on your property, no matter how choked with weeds, by all means use it. Where plants have grown, they're apt to grow again. But there are a number of other factors to consider when choosing a site for the garden:

### Drainage

Drainage is one of the most important factors to consider. Land sloping gently toward the south, southeast or east is ideal. The slope will drain the soil and catch sun's warmth in spring and fall, automatically providing you with a longer growing season. If the rise is a little more than gentle, be sure that rapid drainage does not become too much of a good thing by leading to erosion. Set your rows crosswise on the contour of the hill.

### A Place in the Sun

After drainage, look for sun. If you're on a small suburban lot, put the garden as far as you can from the shadows thrown by your house and your neighbor's. However much you would welcome the shade of a big tree while working in the garden, keep in mind that most vegetables are sun-lovers. Especially keep away from shallow-rooted trees whose roots will quickly horn in on your plot once you've cultivated and enriched it.

### Accessibility

Think of convenience when deciding where to put your vegetable garden. You will save steps if the garden is near as many of the following as possible:

- a source of water
- a toolshed or barn for storage
- a good compost pile location
- the kitchen door (for the convenience of the cook)

## Preparing the Garden

Initial groundbreaking for a new plot, or for restoring a long-dormant one, can be done in several ways, but all of them require planning in advance. You will be off to a better start in spring if you prepare the soil the preceding fall. Remove the larger, woodier vegetation and any large rocks from the land if it has been fallow for long.

Next, the soil must be turned to loosen and aerate it. Digging by hand is the most difficult method but the very best one for the soil. In a small garden, hand-digging is the most sensible way to turn the soil. If you are starting with a lawn, you will first have to remove the sod. To do this, mark the outline of the garden plot by digging straight down into the ground with a spade. You are actually cutting the outline of the plot with the spade. Then slide the spade under the sod and lift it out in pieces. Don't throw the sod away. You can place it upside down in the bottoms of furrows or planting holes, or you can pile it someplace else in layers with like sides together (grass to grass, then roots to roots) and let it stand until it decomposes into a rich soil called "turf loam." When the sod has been removed, you are ready to start digging.

The purpose of digging is to loosen the dark topsoil by breaking up clods and letting in air. Plant roots grow better in loose soil. By turning the soil, richer topsoil can be placed in the root zone and the subsoil can be enriched with additions of compost and fertilizers.

The act of gardening calls for certain tools in addition to those that are the most fundamental—your hands. A few basic tools can make the activities of digging and preparing the soil, fine-tuning the seedbed, planting, and cultivating go smoothly and effectively with less effort on your part. The basic complement of gardening equipment should include (from left to right) a hoe, rake, spade, trowel, shovel, and spading fork. Items like rotary tillers, wheelbarrows, and wheel cultivators are useful too, but generally only in large gardens.

Here is an easy way to dig: first, make a trench the depth of a spade blade, removing and piling up the dirt. Then, dig another trench next to the first one, this time transferring dirt, a small spadeful at a time, into the first trench. In unloading the spade, turn it over and strike it against the side of the trench to break up the dirt. Finally, carry the piled-up dirt from the first trench around to fill the final trench you dig. If your soil is clay, wait until it is fairly dry before digging, plowing, or tilling.

If you are determined to start with a large plot, the soil preparation is going to be more laborious and time consuming.

However, there are still some things you can do to make the job easier. The easiest way to prepare a large garden is to hire someone with a rotary tiller to do it for you. Sometimes they advertise in want ads. Or, you can rent a tiller and do the work yourself.

## Knowing Your Soil

Digging, even if you have the plot plowed for you, is a good way to get acquainted with the soil. If you have dug down a foot, you have probably hit subsoil. It is paler in color than topsoil and full of either sand or clay. Check the topsoil depth. If you find an 8-inch layer, you are lucky indeed.

Look at the wild plants and trees on your property. Burdock, pigweed, lamb's-quarters, clover, and purslane are indicators of good organic content; fennel, sorrel, and chamomile grow on poor

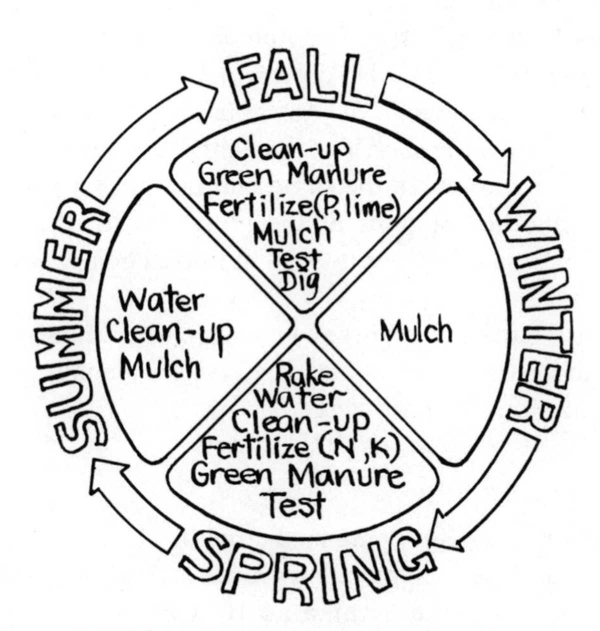


Tending the soil is a year-round process, as shown in this simple cycle. The arrival of each season brings a reminder to take care of certain needs the soil has, in order to ensure its ongoing health.

soil; and scrub oak, wild blueberries, and mountain laurel usually show that the soil is acid.

The texture of subsoil helps to determine how a soil drains. Sand or gravel subsoils may cause plants to wither in drought, but clay drains poorly. The best cure for both overdrainage and underdrainage is to build up a thick topsoil full of organic matter that will bind particles into clusters or soil granules.

## Soil Testing

Once you have investigated soil texture, you would be wise to learn about soil chemistry. Contact your local county extension agent and ask about the soil-testing program run by the U.S. Department of Agriculture. Or you can find a private laboratory where, if you pay for it, you can get an analysis of the organic matter, trace elements, and pesticide content of your soil, along with a basic chemical analysis.

Here's how to take a soil sample for testing: on a dry day in early fall, use a clean trowel to dig a 6-inch-deep hole in the garden. Take a slice from the side of that hole and discard the top and bottom 2 inches of the slice. Put the 2-inch piece that remains into a clean plastic bucket. In the same manner, gather samples from at least four other sections of the plot. Mix up the soil in the bucket, and use a sample of it to fill the small bag provided by the testing service.

When results are reported to you in six weeks or so, your soil-building efforts can begin. Test results usually come with recommendations for fertilizing.

## Acidity and Alkalinity

Acidity and alkalinity are measured in pH units. A pH of 7 is neutral. An acid peat bog can have a pH as low as 3, and an arid alkaline desert soil one as high as 10. Obviously you won't be trying to garden on either of those. But most plants do best where the pH is between 6 and 7. Acid soils can lock up major nutrients.

If you don't have time for a complete soil test this year, at least do a pH test. You will need some blue litmus paper from the drugstore and mud made from your soil sample and clean rainwater or distilled water. Dip three pieces of litmus paper in the mud, wait 10 seconds, and rinse one piece with water. If the paper is pink, your soil is quite acid. Leave the second paper in the mud for 5 more minutes. If it doesn't turn pink before then, your soil is less acid. If, after 15 minutes in the mud, the third paper is still blue, your soil is neutral or alkaline.

If the litmus test or lab report tells you that your soil is acid and needs lime, you can increase pH by one unit by spreading 70 pounds of crushed dolomitic limestone on every 1,000 square feet of loam soil. Clay soil takes 80 pounds of lime, and sandy loam, 50. Spread the lime in the fall. If you have dry wood ashes, use those instead, sprinkling them liberally over the plot. But if you use large quantities of ashes, don't add any phosphorus or potassium fertilizers until you have tilled and tested the soil again. Wood ashes also contribute phosphorus and potassium.

You can amend alkaline soil by digging in oak leaves, pine needles, cottonseed meal, or acid peat moss.

## Building Soil Fertility

Your basic objective is to feed the soil, not necessarily the individual crops that are growing in it. When you incorporate a wide variety of natural materials, you insure that the proper nutrients will always be present in the soil, where they will be released slowly as the plants need them, rather than in a single massive dose, as occurs with chemical fertilizers.

The basic materials that boost and maintain your soil's fertility fall into two categories: organic matter and rock powders. Organic matter encompasses such materials as manure, compost, leaf mold, bone meal, dried blood, and wood ashes. Organic matter supplies the major nutrients—nitrogen, phosphorus, and potassium—in varying amounts, and also plays an important role in making these nutrients available in forms that can be used by plants.

An excellent way to improve soil texture and fertility is to grow green manure crops. If you have a large plot you can incorporate green manuring into your garden planning without having to sacrifice the harvest. Instead of tying up the whole garden all season with a green manure crop, compromise by dividing it into two sections and growing vegetables in one section and a green manure crop in the other. The presence of green manure plants during the growing season can offer an unexpected benefit for vegetables; some green manure crops produce flowers that can lure honeybees into your garden to pollinate the squash and cucumbers. Each year, alternate the crops that grow in each section, so that the vegetable plants can capitalize on the increased fertility of the soil where the green manure crop grew the year before.

Crops in an organic garden get most of their nitrogen as an end product from the decomposition of organic matter caused by soil bacteria and earthworms. In order for the decomposition process to continue, you must add regular infusions of organic matter. However, you must pay attention to the type of material you're adding. If, for example, you were to add some dry plant matter that was high in carbon but low in nitrogen, the bacteria would have to draw upon the nitrogen present in the soil to fuel their breakdown of the carbon material. Although decomposition of this woody material would eventually release nitrogen to the soil as an end product, there would be a temporary deficiency of nitrogen available for plant growth. To counteract this drain on the soil's nitrogen, you should add a high-nitrogen substance such as blood meal or manure whenever you add woody plant matter to the soil.

Organic matter contains considerably less phosphorus than nitrogen, but its value lies more in making soil phosphorus already present from other sources available to growing plants. A soil rich in organic matter is rich in soil bacteria, which secrete acids that promote the breakdown and availability of phosphorus. Without ample organic matter, phosphorus in the soil would be locked up in insoluble compounds.

Organic matter also makes potassium available to plants. Most of the soil's potassium is bound up in mineral form and is therefore unavailable to plants. Some potassium appears in soluble form which plants can use, but there is the danger that it can be quickly leached from the soil before plants can draw upon it. Organic matter helps hold soluble potassium in the root zone, and helps change mineral potassium into a form acceptable to plants as they need it. In short, organic matter helps balance the potassium level in the soil.

Rock powders are substances derived from natural materials which complement the use of organic matter. Commonly used rock powders are phosphate rock, granite dust, and greensand; the first is rich in phosphorus, and the last two are good sources of potassium. All three are available in many garden centers.

By using rock powders, you build the natural phosphorus and potassium reserves in the soil; by working in plenty of organic matter you boost the soil's nitrogen level and insure that the soil nutrients will be readily available in the form that plants can use as they grow.

Soil conditioning should be done in the fall if possible. When you dig your garden in autumn, you give birds a chance to eat the grubs and cutworms that live hidden under grass roots all summer. The freezes and thaws of winter help to further condition the soil. But most important, adding compost and other organic soil-building materials to the garden in fall gives the materials a chance to break down and begin releasing their nutrients into the soil in time for next season's plants.

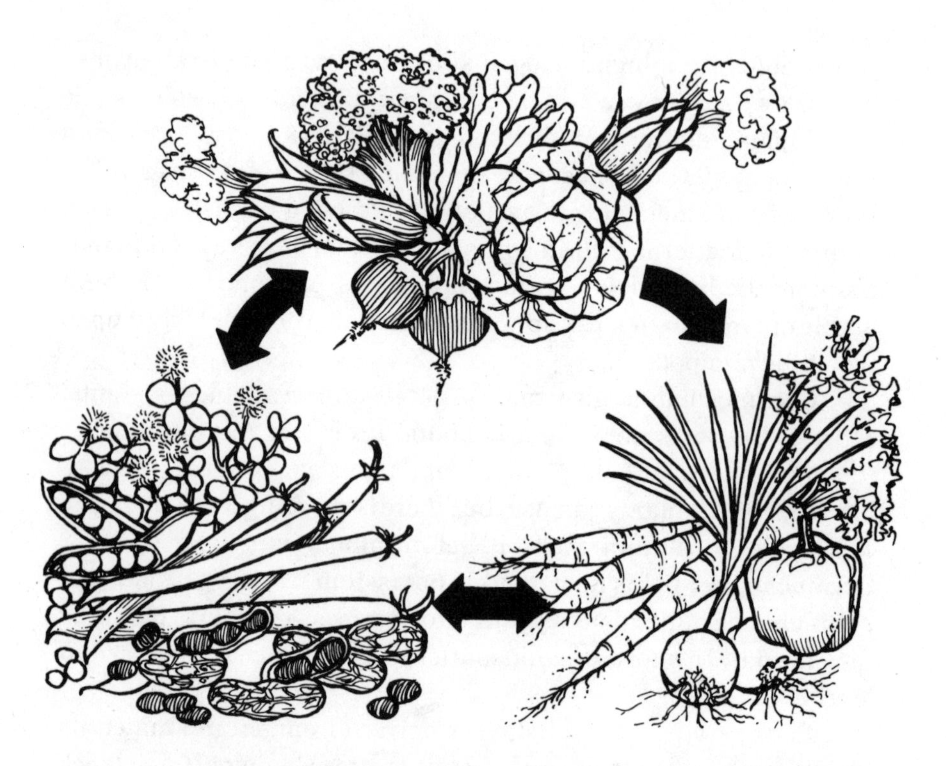

Plants, like people, have different appetites. Some like to consume more than others, and some are even kind enough to replace almost everything they remove from the soil. To keep nutrient levels balanced within your garden, try not to grow the same vegetables (or those with similar appetites) in the same spot year after year. A good rotation cycle to use is to follow heavy feeders (top) by light feeders (lower right) by soil builders (lower left); or, follow heavy feeders by soil builders by light feeders. The only limitation is that heavy feeders should not follow light feeders.

## Composting

Organic gardeners prefer composting to all other methods of returning organic matter to the soil. There are numerous advantages to composting. Because it usually takes place in the presence of oxygen, the decomposition of materials occurs much faster than it does in nature or when a layer of mulch is left on the soil surface to decay slowly. Harmful bacteria and weed seeds are killed by the intense heat generated during composting. Chemically balanced by the manner in which the heap is constructed, finished compost delivers predigested nutrients to soil in forms easily used by plants. It encourages and feeds a large and wide-ranging community of soil organisms and microorganisms which later nourish plants. Incidentally, a properly-made compost pile does not produce odors as it decomposes.

There are almost as many methods of making compost as there are organic gardeners. We will explain two basic methods here: a simple technique that requires several months to produce finished compost, and a quick method that involves more work, but delivers results more quickly.

## The Simple Method

This process takes place in a container or a neat heap. The simplest type of container for making compost is a bottomless cylinder of hardware cloth held together with clips or short pieces of wire. Keep a large pile or second bin full of dry material—leaves, sawdust, ground corncobs, or something similar. If possible, have this material chopped or shredded. Whenever you have a pail of kitchen garbage (except for meat scraps), a basketful of green weeds, a wheelbarrowful of manure or other wet organic materials, scatter them in the compost bin or on the pile and cover them immediately with a thick layer of the dry material. You can continue doing this as you get wet materials. Make sure the pile never dries out completely. If it starts to develop an odor, or whenever you have ambition or need some exercise, "turn" the compost by forking the material out of the bin and back in again. If your compost is in a simple pile, stir it around and fluff it to put some air into it. After the pile has been aging for at least three months, dig underneath and, if the bottom layer feels cool and looks like rich earth, scoop it out and use it.

## The Quick Method

All you need to make fast compost are a 4-foot-high bin that will hold at least 1 cubic yard of material, a good pitchfork, a water source, and some physical strength and energy. Begin by assem-

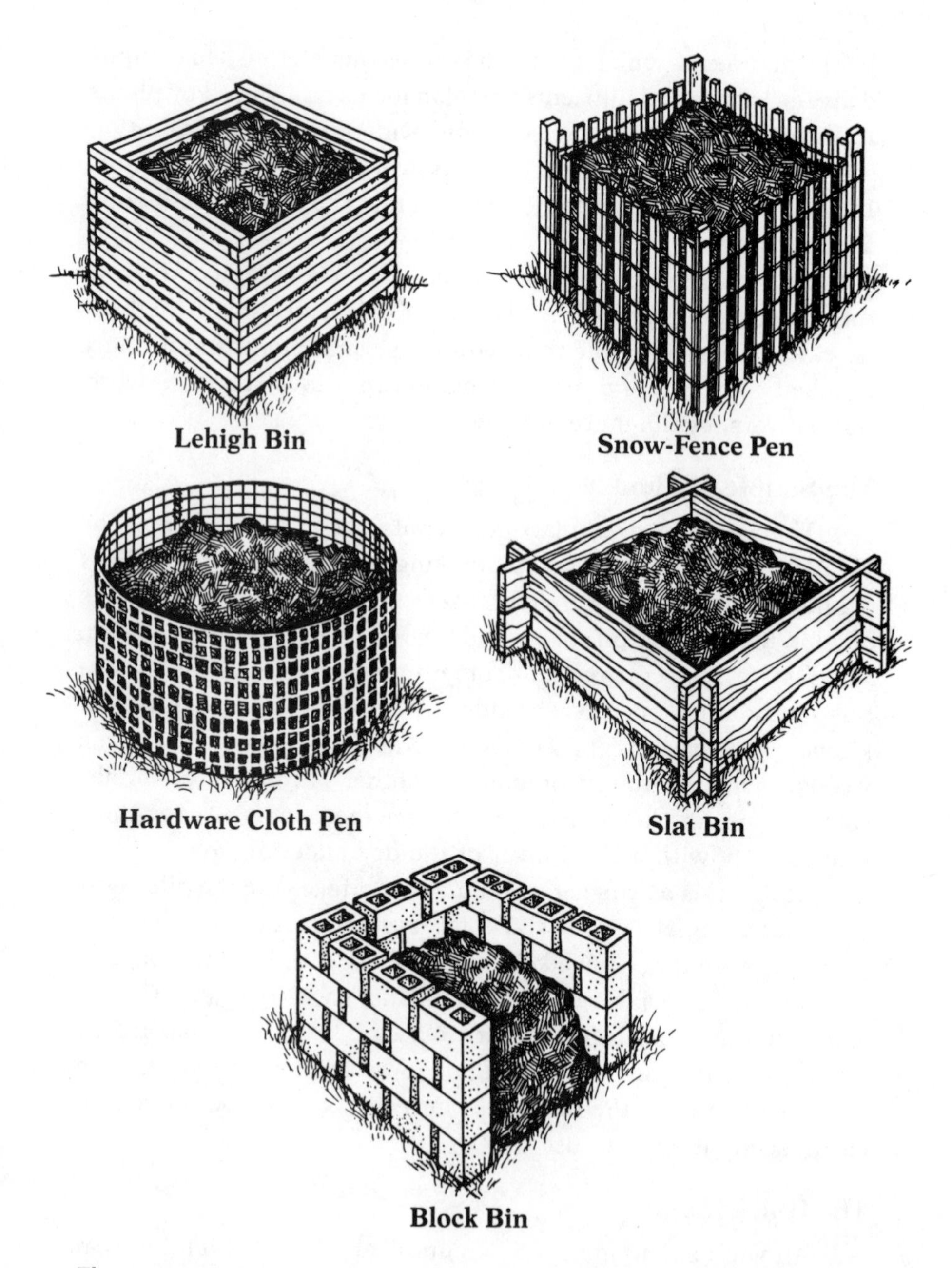


These compost bins and pens are easy to assemble from readily obtainable materials.

bling a large quantity of organic materials. Separate these into two categories: green and/or wet materials; and dry, coarse, woody materials. Chop, cut, or shred all materials into pieces of less than 6 inches.

Fork alternating layers of wet and dry material into the bin, trying to keep the ratio of dry to wet at about 25 to 1. This will mean a 4-inch layer of dry matter, followed by a sprinkling of wetter material. Between layers, and after building the pile to a height of over 3 feet, water or hose down the material until the particles glisten, but not until water runs off. Three days after building the pile, fork out the bin contents, separating the material on the outside from the hotter material inside. Fluff the material as you fork it back in, making sure the outer portions of the old pile end up in the interior of the new pile. It should be mixed up enough to obliterate the layers. Add moisture if necessary to maintain a glistening appearance. Skip a day, and then turn again in the same manner. Skip another day, and fork in and out once more. When the pile has cooled (by about the fourteenth day it will go down to about 110°F), move it out of the bin to age a few weeks. The compost is then ready to use.

This quickly-made compost is not high in nitrogen, but it does wonders for soil structure. Gardeners whose soil is not in top condition would do well to make fast compost at least once a growing season, either in early fall or late spring.

## Using Compost

Dig or till finished compost into the garden in the fall, bury it in trenches close to the root zone, and put it in furrows before you seed and into holes when you set out seedling plants. Use it as a topdressing, sidedressing, or mulch. Favor the heavy-feeding vegetables such as beets and broccoli; legumes can take better care of themselves.

# 2 Planting

The first warm, sunny day in early spring is a powerful stimulus for gardeners. Like iron filings to a magnet, they are drawn outdoors to stand in the garden and contemplate the reawakening happening all around. The packets of seed ordered during the winter have been filed away for what seems like an eternity, and your fingers are itching to get into the soil. But hold on a minute before you go rushing out the door with seed packets in hand. Has your garden plot been plowed or turned since fall? Is the soil dry enough to work?

## Preparing for Planting

Give the soil a "mud test" for moisture both before spring tilling and before planting. Press a clump of mud into a ball in your hand. Squeeze, then release your fingers. If parts of the soft ball fall away, the soil is ready to work. If it all clings together in a wet, sticky mass, it's too early. Wait a week and try the mud test again. Sandy soil, which probably won't even form a mudball, is not harmed by being worked when it's wet. Other soils, however, shouldn't be worked when wet because they will be compressed by your weight and there will be fewer tiny air spaces for plant roots.

If your soil has passed the mud test and the tilling has been done, get out in the garden with a spading fork and break up any clods still remaining where you plan to plant your peas and early salad crops. Rake the top 2 or 3 inches of soil until no big lumps remain. Leave a little roughness so water will penetrate; usually the tiny furrows left by a metal rake are sufficient.

## When to Plant

It's easy to grow enough vegetables to give spring menus a lift, but not all vegetables are suitable for early planting. Good candidates for spring are broccoli, cabbage, carrots, lettuce, onions, peas, radishes, spinach, and turnips. Just make sure the varieties you choose are not only frost-hardy but also early-bearing. WALTHAM broccoli, for example, though it thrives in cool weather, takes 74 days to mature and will bolt in hot summer weather. This variety is best planted to mature in fall. GREEN COMET HYBRID broccoli, on the other hand, matures in only 55 days and can be harvested before hot weather arrives; it's a good choice for spring planting.

Another step in the selection process is to check for special varieties that are generally planted later. For example, although most beans prefer dry, warm soil for germination, bean seeds that are unusually rot-resistant, like those of the variety ROYALTY, can be planted right around the date of the last expected spring frost. The table, "When to Plant Vegetables," will give you an idea of when to plant crops.

### The Last Frost

Your most important guides to planting are the time of your area's last spring frost and its first frost in fall, both of which may vary from year to year. How do you know when to expect the last frost? For one thing, you can learn to read natural "clocks": the last frost in spring usually occurs about the time that the oak trees leaf out. Experienced local gardeners or a county extension agent will be able to give you the average date of the last frost in your area, but remember that you're dealing with estimate, not an immutable law. Frost dates vary according to weather and the environmental conditions in your garden. For instance, if you live in a valley, your last frost will probably be later than it will for your neighbor up

## When to Plant Vegetables

This chart shows some common vegetables grouped according to the approximate times they can be planted and their relative requirements for cool and warm weather.

**Plant in Early Spring**

| Very Hardy (plant 4–6 weeks before frost-free date) | | Hardy (plant 2–4 weeks before frost-free date) | |
|---|---|---|---|
| Broccoli | Potatoes | Beets | Lettuce |
| Cabbage | Radishes | Carrots | Swiss Chard |
| Onions | Spinach | | |
| Peas | Turnips | | |

**Plant in Late Spring or Early Summer**

| Not Hardy (plant on frost-free date) | Somewhat Heat Tolerant (plant 1–2 weeks after frost-free date) | Heat Tolerant (good for summer planting) |
|---|---|---|
| Beans, Snap | Beans, Lima | Beans (all types) |
| Corn | Peppers | Corn |
| Cucumbers | Potatoes, Sweet | New Zealand Spinach |
| New Zealand Spinach | | Squash |
| Squash | | Swiss Chard |
| Tomatoes | | |

**Plant in Late Summer or Early Fall**

| Hardy (plant 6–8 weeks before first fall freeze) | |
|---|---|
| Beets | Spinach |
| Lettuce | Turnips |

the hill (because cold air drains down into low areas much like water does).

## Seeds or Transplants?

Use the table, "Plant Response to Transplanting," as a guide in deciding which crops to seed directly in the garden and which ones to start from transplants. The planting instructions in the chapter "How to Grow 20 Favorite Vegetables" will provide further information. If you live in the North, you may want to use transplants started from seed indoors for long-season crops like cucumbers, head lettuce, onions, peppers, and squash. Of these, deep-rooted plants like cucumbers transplant with the greatest difficulty.

Young plants react to changes in temperature and light conditions and are sensitive to any disturbance of their roots. Even minor damage to root hairs can upset the internal movement of moisture and nutrients. If you are new to gardening, it would be best to use seedlings only for those crops which transplant readily and plant the rest of the garden with seeds. After you've acquired more skill in handling transplants, you can take advantage of the growing time gained in setting out transplants of the more sensitive crops as well.

### Planting from Seed

The first step in sowing seeds is to mark the planting space in the garden according to your plan. For a row garden, use stakes and string to lay out parallel rows. The amount of space between the rows depends on what you're going to plant. (See the chapter "How to Grow 20 Favorite Vegetables" for that information.) Using the taut string as your guide, scratch a shallow line with the corner of your hoe. For beets and peas, which have larger seeds, go a little deeper. Some people lay their hoe on the ground and step on the handle to make a shallow trench.

Although seed packets always specify planting depth, bear in mind that this should always be relative to growing conditions.

## Plant Response to Transplanting

| Transplant Well | |
|---|---|
| Broccoli | Onions |
| Cabbage | Peppers |
| Lettuce | Tomatoes |
| **Transplant with Care** | |
| Beets | Spinach |
| Carrots | Squash, Summer and Winter |
| Cucumbers | Swiss Chard |
| Radishes | Turnips |
| **Do Not Transplant** | |
| Beans | Potatoes |
| Corn | Potatoes, Sweet |
| Peas | |

Plant seeds more shallowly if the soil is heavy or the temperature is low; plant a bit deeper if the soil is light or the weather is hot. Drop large seeds into the furrow one at a time. Small seeds can be mixed with three times their bulk of fine soil or sand before planting, to make them easier to handle. Use the corner of your hoe to draw soil over the row, or sprinkle the seeded row with fine material such as the potting mixture you use for starting seeds indoors. Firm up sandy or organically rich but light soil by pressing with your hands, but leave clay soil unpacked.

Although a distance between plants in a row is suggested for each vegetable in Chapter 5, it's often difficult to achieve consistent spacing between plants when sowing. Not all seeds will germinate, but most beginning gardeners still sow too thickly and then must spend a lot of time thinning plants later on.

## Planting Methods

The most common planting method is row planting, described above. Most vegetables can be planted in rows.

For plants like squash that sprawl all over the ground as they grow, hill planting is the best way to go. To form a hill, dig a 4- or 5-inch-deep hole and fill it with rich compost. Then heap a 4- to 6-inch-high mound of soil on top, and firm the sides. Plant the seeds in the top of the hill according to the space needs of the crop. By planting in hills you'll supply these heavy-feeding plants with ample nutrients, and at the same time you can keep track of where their roots are so they won't be injured when you hoe and weed. Hill planting is the best method for squash, cucumbers, melons, and pumpkins, and can be used for corn.

For planting a large or concentrated area like a wide bed, the technique of broadcast seeding can be used. Broadcasting takes less time and effort than sowing all those seeds individually, especially if the seeds are small. When the seedbed is prepared, just take a handful of seeds and scatter them as evenly as you can over the soil surface. Rake the soil lightly to cover them. If your soil

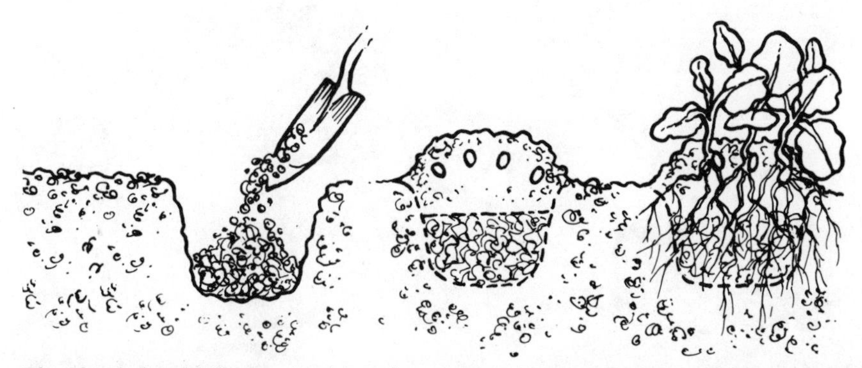

The idea behind hill planting is to plant seeds in a mound of soil slightly higher than the surrounding soil. This "hill" provides a warm germinating area for heat-loving crops like beans, corn, cucumbers, pumpkins, and squash. To nourish heavy-feeding crops, fill the bottom of a 12-inch-deep hole with 8 inches of well-rotted manure or compost, then add 8 inches of soil. Space seeds evenly along the top and sides of the hill. After they have germinated, thin to the strongest seedlings to allow adequate room for growth.

is light, tamp the area gently with the back of a hoe or the bottom of your shoe. Broadcasting is a good planting method for small seeds like those of lettuce or turnips, but you'll have to thin later on, and weeding will be tricky until the plants grow large enough to be distinguishable from the weeds.

## Working with Transplants

Setting out plants instead of seeds is a great time-saver. In areas where the season is short, it may be the only way to get warm-climate crops like tomatoes and cucumbers before fall frosts. There are two ways to get transplants: buy them from a nursery or grow them yourself. Raising seedlings indoors can be tricky, and if you are new to gardening, the safer path is to buy your transplants from a reliable local source. Ask whether the nursery itself grows the plants. Seedlings grown in a small, local nursery will often do better in your garden than plants shipped from a large wholesale grower.

To decrease root shock during transplanting, buy single plants

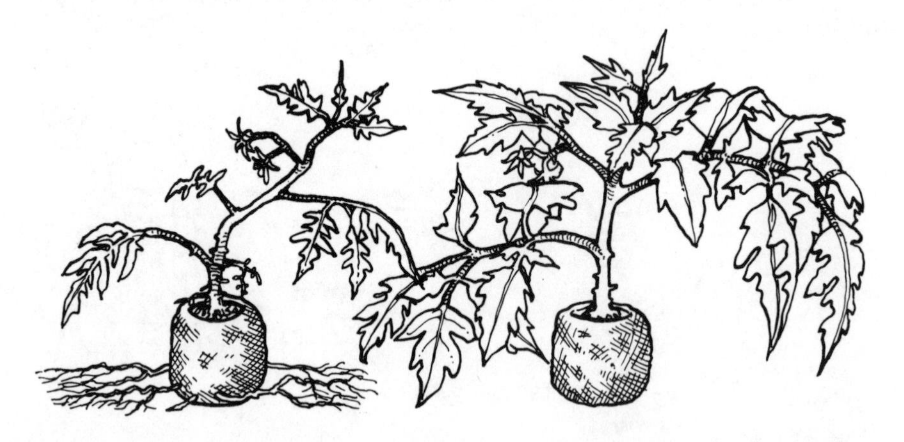

No matter what the price tag says, the seedling on the left is no bargain. When purchasing seedlings to set out in your garden, look for plants like the one on the right that are stocky, not leggy, with healthy green foliage. Give them a thorough going-over to check for pests and signs of disease. Inferior seedlings usually turn out to be inferior plants.

in individual peat pots rather than groups of plants in flats. If individual seedlings aren't available or are too expensive, plants in flats divided into compartments are a better choice than plants in undivided flats.

Check the plants carefully before you buy them. Make sure that the leaves are full and healthy and look for signs of bugs in the soil, along stems, and on the leaves. And remember that bigger is not necessarily better. Shop for sturdy, stocky plants with stout stems and bushy foliage—they will transplant better than tall, lanky plants with long stems. Finally, keep in mind that young plants take to transplanting more readily than older, more established ones. Avoid plants already in bloom.

## Growing Your Own Seedlings

If you're considering growing your own seedlings indoors or in the greenhouse, there are several factors to weigh before making

Despite their dissimilar appearances, these seed-starting containers have several things in common. They allow for good drainage, are spacious enough to accommodate seedlings, and where practical, have been thoroughly cleaned or sterilized. From left to right, the containers shown here include plastic pots, trimmed milk cartons, peat pots, peat discs (compressed and expanded), a cast-off muffin tin, newspaper cuffs, and an egg carton.

up your mind. The advantages of growing your own transplants include:

- It saves money.
- It almost guarantees healthier, disease-free plants.
- More vegetable varieties are available in seed form than in plant form.
- It gives you some gardening to do during the cold, dreary days of February and early March.

The disadvantages include:

- It takes time, trouble, and planning. In northern areas, unless you're willing to spend money on potting soil, you must prepare a planting mixture before the ground freezes solid the fall before.
- It's messy.
- If you fail late in the game—say you wait until the last possible moment to set out your broccoli only to have the plants mowed down by cutworms—you may have trouble locating substitute plants in greenhouses.
- It requires equipment and space. You'll need containers and, for some seedlings, a way of giving heat and light.

## Planting in Flats

Seedlings can be started in practically any container, as long as it has drainage. Wooden "flats" 15 inches long and 4 inches deep, with the bottom fourth covered with sphagnum peat moss, are ideal seed-starting containers. You can buy them or make them at home from scrap lumber. Milk cartons split horizontally work well, too. You can also use plastic foam drinking cups, flowerpots, or individual peat pots.

Put a layer of sand or perlite in the bottom of the flat, then fill the containers almost to the top with a planting medium made of equal parts of peat moss, vermiculite, and sterilized potting soil. Using your finger or a tongue depressor, make tiny furrows, to the depth suggested for the seed. If you are planting more than one kind of seed in a flat, try to choose those with similar germination times. Fill in the furrows and mist the soil to dampen it thoroughly. Water until the soil mix is wet throughout.

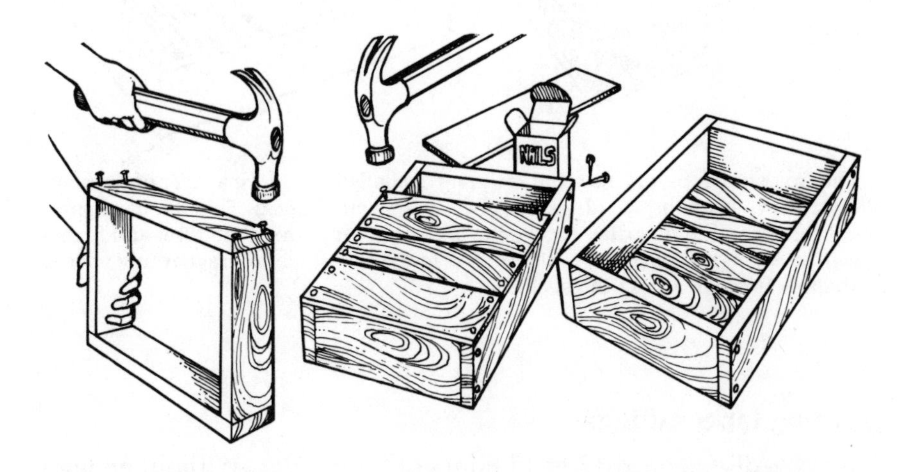


With some scrap lumber and a little know-how, you can build a first-rate seed-starting "flat." Start with two equal sides and two equal ends and nail the frame together. Set the frame down and nail random-width pieces of wood across the bottom, leaving a slight space between adjacent pieces (to promote good drainage). What you've got when you're done is a low-cost, no-frills container ready for seasons of use.

Set the flats (covered, if you like, with plastic, paper, or glass to keep the humidity high) in a warm, dimly lighted place—next to a wood stove, furnace, or radiator, or in a warm closet, for example. Remove the cover at least once a day to allow some air circulation and to check for signs of fungus and for germination. If you see either spots of white (fungus) or green (plants) remove the cover and take the flat to a sunny spot. If you have a severe infestation of fungus, you may have to start over, although sometimes scraping it off and exposing the soil to sunshine will give the seedlings another chance. If germination has taken place, more seedlings will continue to emerge in sunlight. If they are too thick, pluck some out with tweezers or snip them off at soil level with nail scissors.

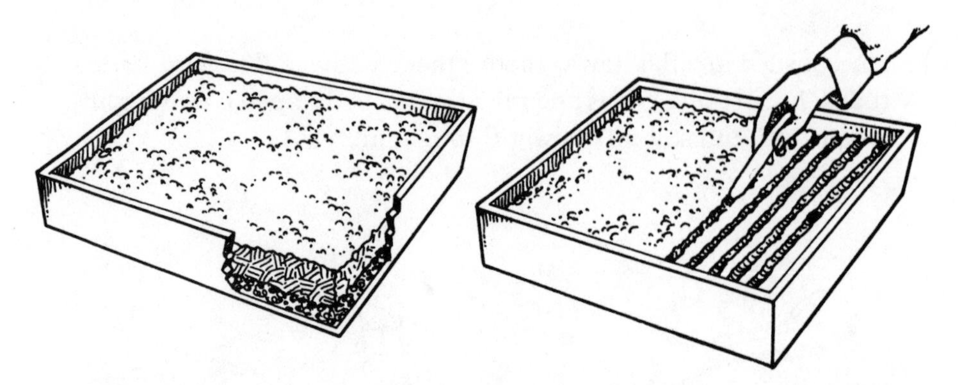

The first step in preparing a flat for planting calls for adding a layer of drainage material, to keep the seedlings from becoming waterlogged. This flat has a layer of coarse sand and perlite. Next pour in the growing mix, then level and firm it gently. A tongue depressor is the perfect tool with which to draw evenly spaced shallow furrows across the surface.

## Caring for Seedlings

Seedlings need 12 to 15 hours of light a day. Without enough light, young plants develop tall, weak stems, sparse leaves, and a severe tilt toward whatever light they manage to find. An ideal way to supply enough light is to combine a bright location like a solar greenhouse, south-facing windowsill, or slightly heated glassed-in porch with a bank of fluorescent lights that are turned on at dusk, then off in time to allow five hours of total darkness. You can also raise healthy seedlings under nothing but fluorescent lights. Use a daylight tube or a combination of warm-white and cool-white tubes. The foliage should be no farther than 4 inches from the lights.

Roots need air to grow. That's why it is essential to use a loose planting medium that won't waterlog. After germination, water sparingly, when the soil is dry to the touch. Use lukewarm water to minimize shock.

When seedlings have developed their first set of true leaves (usually the second leaves the plant develops, but the first to bear the recognizable characteristics of the plant's mature foliage), you'll need to provide extra nourishment for the growing plants. Feed them occasionally with fish emulsion or other mild liquid

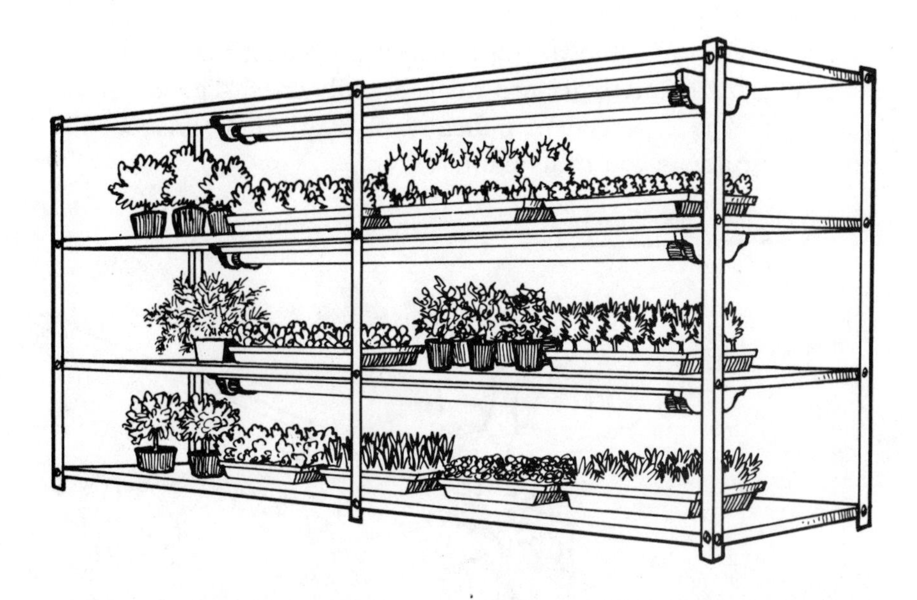

Young plants need 12 to 15 hours of light a day to get off to a good start. You can make sure they get ample light by constructing a light garden like this in your basement or spare room. The triple-decker arrangement uses space efficiently and provides plenty of room for trays and pots of vegetable seedlings destined for the garden.

fertilizer, and increase the light so the plants will be able to put the extra nutrients to good use.

Seedlings kept too warm often put forth spindly, unhealthy growth and, like hothouse plants, get whomped when they finally meet the cold, cruel world. A good temperature during the day is 60° to 65°F, with a 5- to 10-degree drop at night. To keep windowsill plants from getting too cold on frosty winter evenings, cover windowpanes at night with newspapers or a screen.

## Transplanting Indoors

Some vegetable seedlings should be transplanted at least once before they are planted outdoors so they develop more and healthier feeder roots. These include seedlings of cabbage, lettuce, onions, peppers, and tomatoes. Transplanting most of these vegetables up to four times benefits roots and makes the tops more compact and sturdy. It also allows you to select the fittest plants at each stage of growth and to discard weak, spindly ones. A good, strong root system is an excellent indicator of plant vigor; inspect the roots carefully as you transplant.

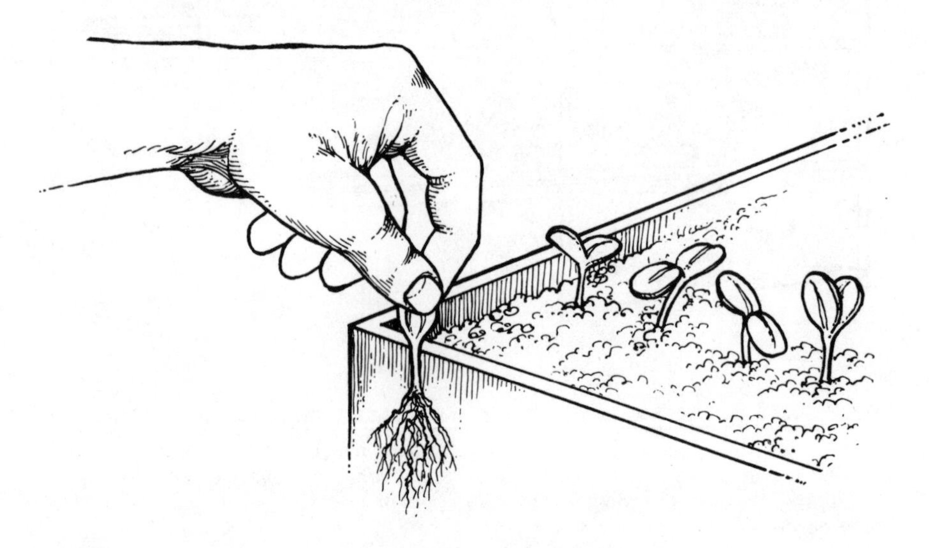

When transplanting, treat the seedlings tenderly as you transfer them to a larger container. Lift them by their leaves to keep from crushing the fragile stems. Try to handle only the tips of the leaves to avoid bruising the growing point that lies at the base of the leaves where they join the stem.

Lift the transplant out of its flat with a spoon, sharpened stick, or tongue depressor. Set it on a damp mound of soil-covered peat moss in a pot or another flat. Water it in, and draw the soil around the stem more than an inch higher than it was before. At each transplanting use a richer mixture and wider spacing.

If tomato or pepper seedlings seem leggy, pinch off the tops to get a stockier plant base. Shear back onions and lettuce, but leave members of the cabbage family alone.

Vegetables that transplant poorly, such as cucumbers, squash, and many root crops, should be started in individual pots or spaced widely in flats to begin with.

## Hardening-Off

Allow seedlings at least a week of gradually increasing exposure to outdoor conditions of light, temperature, and air circulation before you plant them in the garden. Before "hardening-off" the young plants in this way, withhold water and fertilizer for a few

### Transplanting Guide for Seedlings

| Vegetable | Weeks to Transplant Size (from time of sowing) | When to Plant Out |
|---|---|---|
| Beets | 4 | 4 weeks before frost-free date |
| Broccoli | 6-8 | 4 weeks before frost-free date to 3 weeks after |
| Cabbage | 6-8 | 5 weeks before frost-free date to 3 weeks after |
| Carrots | 5-6 | 4 weeks before frost-free date |
| Cucumbers | 4 | frost-free date to 8 weeks after |
| Lettuce | 4-6 | 2 weeks before frost-free date to 3 weeks after |
| Onions | 4-6 | 6 weeks before frost-free date to 2 weeks after |
| Peppers | 6-8 | 1-2 weeks after frost-free date |
| Radishes | 3 | 6 weeks before frost-free date |
| Spinach | 4-6 | 3-6 weeks before frost-free date |
| Squash | 4 | on or after frost-free date |
| Tomatoes | 6-10 | on or after frost-free date |
| Turnips | 3-4 | 4 weeks before frost-free date |

Transferring seedlings from their containers to the garden—called "planting out"—is a simple procedure when you follow these basic steps. Carefully slip seedlings out of their containers into your cupped hand (upper left); a gentle tap with a trowel on the bottom of the pot can coax a reluctant root ball out. The holes should be ready and waiting for the unpotted seedlings (lower left). Make sure the holes are roomy enough to accommodate the root ball, and add extra soil as needed to fill in the spaces between the roots and the sides of the hole. Firm the soil around each seedling and create a slight depression around the stem. Finish off with a draft of starter solution (upper right) that will firm the soil some more and give the seedlings a gentle nutrient boost.

days. Then put the plants outdoors for increasing periods, working up gradually from three hours outside to a whole day.

## Planting Out

Transplant potted seedlings you have purchased to the garden as soon as possible after you get them home. If you must wait, give them shade and water until planting. A week before transplanting your homegrown seedlings, plants in flats need to be "blocked" to separate their roots. Use a sharp knife to cut the dry soil in the flats into cubes, each one containing a plant. Blocking several days before transplanting gives the roots a chance to heal and reestablish themselves.

Homegrown plants can be set out according to the "Transplanting Guide for Seedlings" table. Don't transplant on hot, sunny days. If the weather is warm, early evening is a good time; seedlings won't have to face hot sun and new conditions at the same time. A cloudy day is ideal.

Loosen the soil in the bed or row. An hour before transplanting, dampen the soil in the pots or flats, but don't get it too muddy. Dig a hole in the bed or row for each plant, and partly fill each hole with compost. When setting the pot or soil cube in the ground, make sure the plant roots are in firm contact with the soil in the hole. Water well. Check to see that the seedling is set the proper distance into the soil. (See Chapter 5 for planting depth information.) Shade your seedlings for a few days, a week if it's hot. You can use cloth or paper covers, baskets, cloth stretched over bent wire, or screens of burlap or cheesecloth—anything that will block the sun but not the air.

# 3

# Garden Care through the Season

Maturing plants have the same basic requirements as seedlings: light, air, moisture, and nourishment. It's up to you to meet these needs as the growing season progresses. Well-planned gardens seldom lack sunlight, so we will not address light in this chapter. Instead, we'll look at air, moisture, and nourishment, and also weeding and pest control.

## Aerating the Soil

It's easy to assume that plants get all the air they need from the atmosphere around them. But there must be air in the soil, too. Roots need the air held between soil particles to help absorb nutrients. To achieve good soil-air circulation, you need loose, crumbly soil with good drainage (air can't fill pores occupied by water). Cultivation with a hoe or cultivator is an excellent way to aerate soil. Mulching the garden helps create a soil environment that is hospitable to earthworms, whose tunneling aerates soil as well.

If your soil is base, take time either to hoe or cultivate between rows, where the soil tends to crust. In addition to aerating soil, cultivation helps keep weeds down, and brings insects and their eggs to the soil surface to die. Be careful not to hoe too close to plants or you may injure shallow feeder roots.

## Watering

The object of watering is to keep a constant supply of moisture available to plant roots. There are numerous watering methods, depending on the size of your garden and your personal preference. Your goal should be to direct the water to the plant roots in the ground, not to the foliage or the flowers. If the garden is small, you can water each plant individually from a sprinkling can or bucket. To provide slow, steady watering for plants that are moisture-lovers, you can make a reservoir from a plastic milk jug. Punch holes in the cap, fill the jug with water and place it upside down next to the plant. Water will trickle slowly from the holes in the cap. For larger gardens and bigger budgets, there are trickle hoses and drip irrigation systems that do the same thing on a larger scale. This slow, even watering is very beneficial because the water

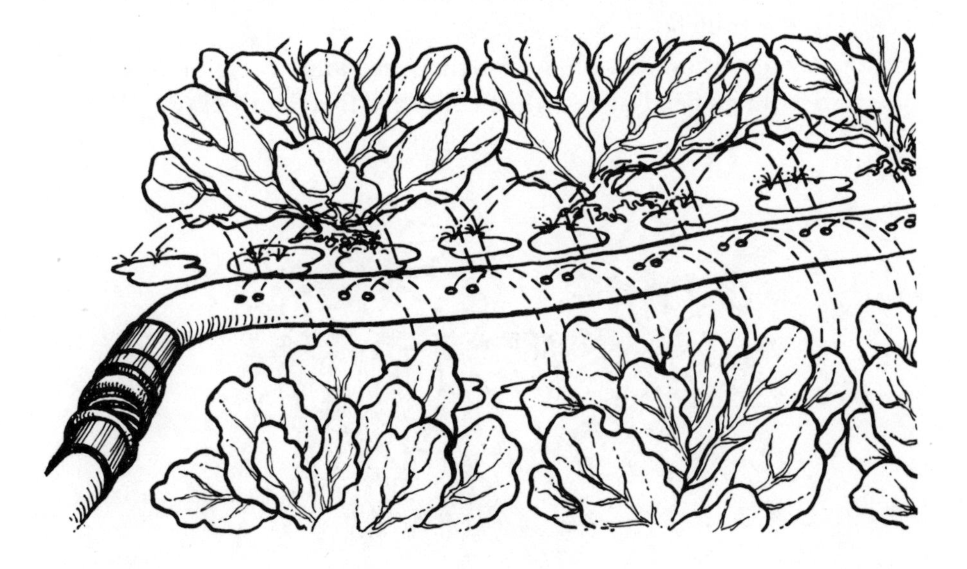

One of the most efficient means of watering the garden is with a special plastic soaker hose that delivers the water right where it is needed, and at a slow and steady rate so that it is absorbed by the soil rather than lost to the air. The hose is hooked up to a regular garden hose which connects to a spigot; this avoids wasting water if the perforated hose has to cross areas other than the garden to reach the water source.

has a chance to soak down deeply to plant roots, and no water is wasted in runoff.

During dry weather, start watering when the soil network becomes thin. If the ground is dry 3 inches below the surface, plants will droop at midday and pick up only slightly in the evening. Dehydrated plants are prone to diseases and insects.

Here are some rules for efficient watering:

- To prevent evaporation and leaf burn, water early or late in the day, never at midday.
- Apply water as close to the plants as possible, taking care not to uproot seedlings.
- Water very slowly and continuously with a drip or trickle system, or else soak the ground as deep as 4 feet once every ten days to two weeks in dry weather.
- Use a fine-spray sprinkler, or use a perforated soaker hose under mulch.
- Pay attention to plants' growth stages and seasonal needs. Seedlings need water when soil dries to a depth of only 1½ to 2 inches.

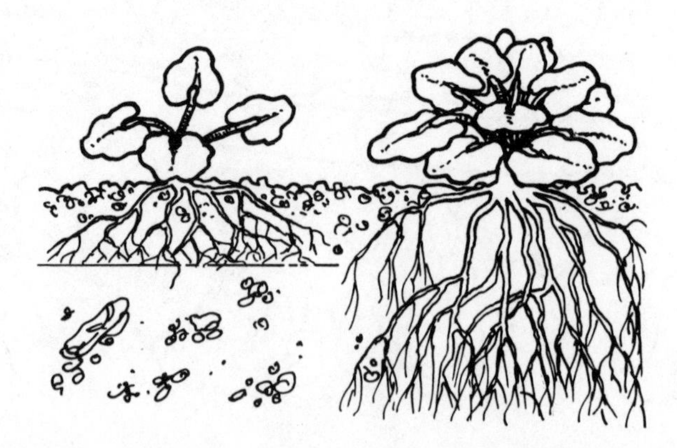

Deep watering encourages deep roots. A plant with a deep root system like the one on the right is better equipped to survive droughts than the shallow-rooted one on the left, since it can reach down to where the moisture level is more stable. A deeper root system also enables the plant to pull up nutrients from a greater area. Encourage deep root systems to form by giving good, thorough soakings at infrequent intervals, instead of dribbling a little water on the soil surface every day.

- Custom-water individual vegetables. (See the guidelines in the chapter, "How to Grow 20 Favorite Vegetables.")
- Remember that overwatering is damaging. Roots grow stronger, deeper, and more efficient if forced to penetrate mineral-rich subsoil in search of moisture.

## Fertilizing during the Season

Many crops, especially heavy feeders, benefit from booster feedings as they grow. A midseason feeding is a good idea, and if your soil is not in top condition, fertilize as soon as young plants are established, too. These supplemental feedings can be given in a number of ways. First, you can topdress or sidedress with compost. To topdress, just spread compost a few inches deep in a ring around the stem of a plant. Sidedressing is done by spreading compost along a garden row, beside the plants. These techniques make nutrients available to plants gradually as rainwater leaches the compost down into the soil near the roots.

A more convenient way to fertilize during the season is to water plants with seaweed solution, fish emulsion (both of which

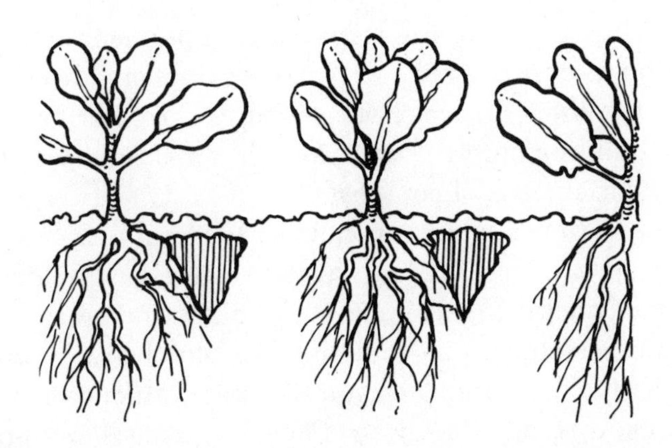

Sidedressing is a simple technique to make midseason feedings of materials like rotten manure, compost, or bone meal. Dig a furrow alongside the plant or group of plants to be fed, add the material, then draw a layer of soil over the furrow with a hoe.

can be found in garden centers), or compost water. To make compost water, fill a burlap sack with compost and suspend it in a drum or barrel full of water. Let the "teabag" steep until the water has turned a rich, brown color.

When your soil is in peak condition and rich in organic matter, feedings are usually unnecessary for all but the most demanding crops.

## Weed Control

Weeds can harbor insects and spread diseases to cultivated plants of their species and rob growing crops of nutrients.

The most important—and the easiest—weed prevention method is to cultivate three days after planting seeds. Pull a steel rake gently across the rows, uprooting germinating weeds. Rake again, more carefully, once your crops are up. Hand-pull weeds growing near crop plants, and as the weather warms, hoe to uproot large weeds and smother small ones.

## Mulching

Weeds also need light to grow. That's why a layer of mulch controls them so well. It also traps water vapor near plants, slows evaporation, prevents wind and water erosion, stabilizes soil temperature by holding heat in or out, nurtures earthworms, and adds structure-building humus to soil. Mulch is a supplement to, not a substitute for, the use of compost.

A wide variety of materials can be used for mulch. Some of the most popular are spoiled hay, straw, seaweed, shredded leaves (unshredded ones compact into a soggy mat), and grass clippings. Add a nitrogen-containing material like blood meal, cottonseed meal, or animal manure when you till under mulches of sawdust, wood chips, or ground corncobs. These materials use up lots of nitrogen as they decay. Man-made materials such as newspaper and black plastic also make effective mulches, but they must be removed at the end of the season.

A 6-inch layer of loose straw or hay, or a 2-inch layer of finer material, such as buckwheat hulls, will stifle weeds. Add an inch of mulch for every inch of rain that is deficient during a dry spell. Push the mulch tightly around the plants. Till organic mulches under in fall to add organic matter to the soil.

## Pest and Disease Control

A healthy garden with sturdy, well-nourished plants and fertile soil is the best insurance against disease and pests. Well-planned crop rotations and a weed-free garden are important preventive measures. But sometimes, no matter how careful you are, harmful insects, infections, bacteria, and fungi strike your vegetables with a vengeance. When trouble strikes, learning to identify the cause of the problem should be your first objective; finding a solution, the second. It helps to take a scientific approach to the world of insects and disease. Perceive your garden as a kind of outdoor laboratory where you can study the good and the bad in nature, the creative and destructive forces at work.

### Know Your Enemy

If you don't know a curculio from a curlew, you're in trouble. Kill the wrong one and you lose a friend and a good part of your crop. A garden is no place to be trigger-happy. Organic methods of controlling insects are based on balancing the total garden environment so that all forms of life, including insect predators and prey, are kept in balance.

The first line of defense is healthy soil and diverse and healthy plants. Insect pests and diseases function to cull weak and poorly nourished plants. Not all insects are pests. Many protect plants and aid in pollination. Whenever you introduce plants to the soil, insects naturally follow, which keeps things in balance. Because insects are part of the garden, you should understand them as well as you do your vegetables.

You can learn to live with a few bugs, and so can your plants.

Most vegetables can withstand some insect damage without a loss of yield at harvest time. You can also time plantings to avoid certain insects, and plant varieties that are insect-resistant.

## Know Your Enemy's Enemies

Birds are major allies in any battle against the bugs. Encourage them by all means. Build houses for the urban types like wrens, and plant the shrubs and trees birds prefer. Accept, as a protection fee, the loss of some of your uncovered garden fruit.

Birds with long, straight or curved bills or with short, whiskered beaks will eat only insects, but birds with fine, sharp bills eat both seeds and insects. Birds need water and nest materials, and they respond to winter feedings of suet or dried fruit, nuts, and seeds. A hedgerow or other planting of low-growing native bushes and trees attractive to birds near the garden will encourage them to visit.

Toads, also voracious insect-eaters, can be encouraged to take up residence in your garden by providing water and shelter in the form of upside-down clay flowerpots with holes chipped out of the sides for entrance.

Beneficial predatory insects also need encouragement. You can order through the mail ladybird beetles (ladybugs), praying mantids, lacewings, predatory mites, *Trichogramma* wasps, and some other wasps with parasitic larvae. Ladybugs and lacewings will gobble up aphids and mealybugs. Female wasps lay eggs inside the eggs of harmful insects, or inside their larvae; the wasps' parasitic offspring eat their way out, destroying their hosts in the process. Instructions for introducing these beneficial insects into the garden are shipped with the insects or eggs. Later, you may be able to learn to raise your own. Beginning gardeners often order predators without first knowing if the garden will supply enough food for them, and so they lose them quickly. It is always a good idea to plant or encourage alternative food sources for insect predators. For ladybugs, a commercially available food called "wheast" can be purchased, and they also like pollen and nectar from plants. The lacewing is attracted to night-blooming crape myrtle and, like predatory wasps, to flower nectar.

Other beneficial insects include spiders, ambush bugs, assassin bugs, firefly larvae, damselflies, yellow jackets, tachinid flies, and other flies that parasitize caterpillars, borers, and various beetles. Get to know these insect friends. A beneficial bigeyed bug looks very much like a harmful leafhopper; only patient observation will differentiate the two.

## The Organic Arsenal

As an organic gardener, you have many defenses available against harmful insects. Begin by cultivating your soil to expose and dry out or freeze larvae and by rotating your crops to improve plant health and foil overwintering pests. Whenever possible, handpick pests from your plants. Equip your garden with insect barriers and traps, such as paper plant collars to thwart cutworms and sticky yellow boards to catch whiteflies. Plant resistant vegetable varieties that insects will find unappetizing or difficult to eat and digest.

Commercially available insecticidal soaps can be mixed with water and sprayed on plants to kill pests. There are natural sprays, and powders which repel or sicken insects, and irritating substances, such as ashes, diatomaceous earth, and limestone, which injure soft-bodied bugs and slugs. One of the most advanced techniques of pest control is the introduction of insect diseases into the garden. Among these, *Bacillus thuringiensis* (sold as Dipel or Thuricide) is the best known. It works on many insects that pass through a caterpillar stage by disrupting and paralyzing their inner cellular structure. Another pathogen, milky spore disease, *Bacillus popilliae,* is widely used to control Japanese beetles (available through the mail and at garden centers).

Botanical insecticides like rotenone, pyrethrum, quassia, derris, sabadilla, and tobacco dusts and sprays, though derived from plants, should be used only as a last resort. Although these materials break down quickly, they can be toxic when first applied and must be handled with care. If you decide to use these products to get rid of a particularly stubborn infestation, follow the package directions carefully and always keep these poisons away from children and pets.

## Plant Diseases

Few beginning gardeners are plant doctors, and amateur diagnosis is always tricky. A diseased plant often resembles one that is infested by insects or deficient in nutrients. Healthy soil, good growing conditions, and keeping the garden free of weeds and other debris are the best defenses against disease. Careful crop rotation is important, and regular cultivation can help reduce soilborne pathogens. Whenever possible, plant disease-resistant varieties. Also refrain from working around plants when they are wet. Never smoke in the garden (tobacco mosaic virus is transmitted to tomatoes and other crops through cigarette smoke), and if you are a smoker, always wash your hands thoroughly before going into the garden.

When disease does attack, pull up and destroy the affected plants immediately to keep it from spreading. Do not put the plants on the compost pile—seal them in plastic trash bags for disposal.

# 4

# Harvest and Storage

Thanksgiving hymns and agricultural lore may lead gardeners to believe that harvest is a one-time joyous occasion. Not so. The more successful the garden, the more continuous the crop. If you've planned your garden well, you'll be picking delicious fresh vegetables all summer and on into fall.

Picking vegetables at their peak demands regular surveillance. With some vegetables, bigger is *not* better. Don't let vegetables get too mature before picking them. The pleasant experience of biting into a just-ripe red beet or a crunchy snap bean is one of the special rewards of gardening. The taste and texture of homegrown vegetables picked when young is unsurpassed by even the most expensive supermarket produce. And the nutritional value of most vegetables is highest before they are fully mature.

Here are some tips on how to pick some common garden vegetables (see the next chapter for further details):

- **Lima beans:** Hold pods up to the light or squeeze them gently to determine if they are full. Pick limas when beans have developed inside the pods, but before they begin to ripen. The insides of the pods turn white during the period when the seeds are ripening.
- **Snap beans:** Pick when the pods are full-sized (about 4 inches long for many varieties) but when seeds are still small in the pods. This is usually two to three weeks after

the first flowers appear. Never pick snap beans when wet—they can develop brown discolorations called rust. Harvest all beans often to keep the plants producing.

- **Beets:** Both sugar content and toughness increase as beets age. For crispness and best flavor, start digging or pulling them when they reach golf ball size.
- **Broccoli:** Pick after the central head stops growing, but before buds turn lighter green, spread, and begin to open. Cut the head with a sharp knife, leaving a long stem.
- **Cabbage:** Cut young, tight heads. Give plants a twist, moving both head and roots a half-turn, to curtail growth if heads are mature but not yet needed; this will also prevent tight heads from splitting.
- **Carrots:** At ½ to 1 inch in diameter, carrots are ready for table use. For storage, dig them at 1½ inches in diameter, when the skins and cores are a bit tougher. Cut back the tops of carrots left to winter over in the garden under a layer of mulch to 1 inch above the ground.
- **Corn:** Pick in late afternoon while the dinner kettle is boiling. Twist or cut the ears from the stalk, husk, and cook them immediately. The natural sugars in corn turn quickly to starch after harvest. A milky liquid in the kernels is a sign of ripeness; test for it by puncturing a kernel with your fingernail.
- **Cucumbers:** Pick when they are dark green and 3 to 4 inches long. Harvest oriental varieties when they are longer and pickling cucumbers when they are shorter. Picking encourages continued production.
- **Onions:** The stage for harvesting depends on the use. Pull bunching onions as needed when 8 inches tall; pull onions for storage one week after tops have fallen at the end of the season. Let them cure in the open air for a few days before storing.
- **Peas:** Taste for crispness and flavor to determine the correct seed stage for your variety and your taste preference. Hold pods up to the light to see how full they are. Harvest pods from the bottom of the plant first. Any overmature, leathery

pods should be picked and dried or discarded. Complete picking encourages further production. Pick snow peas when pods are still flat, and snap peas before the seeds inside the pods swell enough to touch one another.

- **Peppers:** Cut, don't pull, peppers from plants when they are full size and deep green or red in color. Frequent cutting encourages production.
- **Potatoes:** Dig early potatoes after blossoms have formed. Dig potatoes for storage after the vines have died back, but before frost. Handle carefully. If troubled by pests, harvest soon after the vine yellows.
- **Sweet potatoes:** The bottom leaves start to yellow slightly when the tubers are ripe, but this seldom happens in northern gardens. Dig just before the first frost. Cure before storing.
- **Salad vegetables:** Spinach can be cut or pulled when the outer leaves are 6 to 10 inches tall. Leaf lettuce can be picked as soon as the leaves are a few inches long. If using as "cut-and-come-again" crops, harvest individual leaves from the outside, picking often to encourage new growth; otherwise, cut off the entire plant at ground level. Head lettuce should be cut when heads feel firm, or, for loose-headed types like Boston, when the head has formed in the center of the plant. When possible, harvest all leafy crops early in the morning, before respiration begins in the plant; this will keep the leaves from wilting after you pick them. Refrigerate immediately.
- **Squash:** Pick summer squash anytime after they're a few inches long. To test for ripeness, press the skin with your thumb; it should make an imprint. Younger summer squash have more flavor and smaller seeds than larger ones, and early cutting promotes production. Winter squash is ready to harvest when your thumb doesn't make an imprint in the skin. Cut the squash off the vine; don't twist it.
- **Tomatoes:** Pick when fruit is bright red (or yellow) and barely soft.
- **Turnips:** Begin pulling spring turnips when they are 2 inches

in diameter. Overmature storage turnips are sometimes woody, but exposure to a light frost can improve the flavor.

Most vegetables — except the leafy ones — should be harvested late in the day. Vitamin C content is highest then. Vegetables ripened in sunny weather will be the best nutritionally.

## Storing the Harvest

Before learning about ways to process food for preserving, investigate all the methods of fresh food storage. Fresh food is often better nutritionally, and usually is cheaper to store and better tasting. Some vegetables can be stored fresh up to several months if kept under the right conditions. For our purposes, storage conditions can be divided into five basic types: warm and dry (50° to 55°F, 60 to 70 percent humidity), cool and dry (35° to 40°F, 60 to 70 percent humidity), cold and moist (32° to 40°F, 80 to 90 percent

*(continued on page 52)*

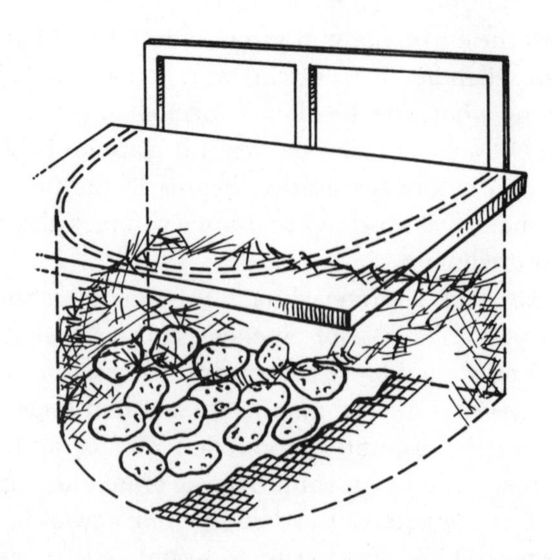

Don't overlook a window well as a possible cold-storage area in your home. Rodentproof it by laying a piece of hardware cloth on the bottom, and weatherproof it by setting a board across the top, followed by a waterproof covering.

Photo 1: For tomatoes that ripen early, use high-phosphorus fertilizer around the base of the plants. Good companion plants are complementary-flavored herbs like basil, rosemary, oregano, marjoram, tarragon, and thyme.

Photo 2: Snow peas can be eaten, pods and all, before the peas fully develop inside.

Photo 3: Repel squash bugs by growing radishes, marigolds, or nasturtiums near your zucchini bushes. Don't overfertilize when planting zucchini or harvest could be delayed by up to two weeks.

Photo 4: Planting corn after the spring ground has thoroughly warmed helps avoid damage from corn borers.

Photo 5: Peas are ready for picking about three weeks after the blossoms appear. Look for pods that are low on the plant because they'll be ready first.

Photo 6: To avoid bitter-tasting carrots, give them an even supply of water and nutrients, particularly potassium, during the growing season.

Photo 7: Whole onions should not be refrigerated. Instead, keep them in a cool, well-ventilated place until you're ready to use them.

Photo 8: Pick peppers when they are firm and full-sized and any color. Most varieties turn from green to red when mature, but some turn yellow and others, a chocolate color.

Photo 9: Don't plant potatoes where they grew last year. Instead, rotate crops—it's best to follow potatoes with legumes. Spacing potatoes close together suppresses weeds. Stored at about 40°F in a dark, humid place, your summer crop of potatoes should last until the following spring.

Photo 10: Try mixing vegetables and flowers in your yard to create an "edible landscape."

## How to Store Fresh Vegetables

| Vegetable | Conditions | Location |
|---|---|---|
| Beets | Cold (32°-40°F) Very moist (90-95% humidity) | In garden under mulch. |
| Cabbage | Cold (32°-40°F) Moist (80-90% humidity) | Store with head down and roots up in buried cache, covered with straw or leaves. |
| Carrots | Cold (32°-40°F) Very moist (90-95% humidity) | In garden under mulch. In root cellar or cold pit surrounded by moist sand or peat moss. |
| Onions | Cool (35°-40°F) Dry (60-70% humidity) | After curing, store in bins or string bags in attic or other cool, dry place. |
| Potatoes | Cold (32°-40°F) Moist (80-90% humidity) | In buried caches where no light can penetrate. |
| Potatoes, Sweet | Warm (50°-55°F) Moist (80-85% humidity) | After curing, store in a warm, well-ventilated but humid room. |
| Radishes (winter types) | Cold (32°-40°F) Very moist (90-95% humidity) | In garden under mulch. |
| Squash, Winter | Warm (50°-55°F) Dry (60-70% humidity) | In an unheated room. Don't let them touch one another. |
| Tomatoes (ripening) | Warm (50°-55°F) Dry (60-70% humidity) | Green mature tomatoes will ripen in 4-6 weeks in a warm, moderately humid place. |
| Turnips | Cold (32°-40°F) Very moist (90-95% humidity) | In garden under mulch. |

humidity), cold and very moist (32° to 40°F, 90 to 95 percent humidity), and warm and moist (50° to 55°F, 80 to 85 percent humidity). See the table on page 51, "How to Store Fresh Vegetables," for storage suggestions for various vegetables.

Potential storage locations include right in the garden under a thick layer of insulating mulch, buried caches (barrels or mounds), root cellars, unheated pantries or basements, unused rooms, attics, crawl spaces, stair and window wells, and outbuildings. Just be sure to match the storage conditions needed by the vegetable with a location that supplies that type of environment.

## Preserving

There are three common ways to preserve food at home—canning, freezing, and drying. Each method has good points and drawbacks. Freezing is quick and easy but uses costly electricity; canning costs little but is hot, heavy work; drying is inexpensive but takes a lot of time and isn't appropriate for all foods. Time, money, and taste are three factors you will want to consider when deciding how to preserve your vegetables.

According to a study done at Cornell University, the total time needed for canning 8 quarts of beans (one preserving kettle load) was 3.2 hours. Freezing the same quantity (16 pounds) took 2 hours, and drying time was 10.8 hours.

Canning foods costs 4¼ cents per pound, according to a study conducted some years ago by *Rodale's Organic Gardening* magazine. That translates to 8½ cents per quart jar. Drying the same amount of food costs 6 cents, and freezing it costs 21 cents, if the food is stored frozen for a full year.

Many gardeners and cooks find that a mixed storage system meets their needs best. Frozen foods taste closest to fresh and can be steamed or stir-fried to retain a crisp texture. Canning lends itself to food that you would normally cook thoroughly, like stewed tomatoes. High-acid foods are safest for canning. Dried foods have the greatest limitations. For example, dried string beans work well in stews, but aren't very appealing as a side dish.

Your choice of preserving methods will also depend on your answers to the following questions:

- Do you already own a freezer or would you have to buy one?
- Is your climate warm enough to grow something dark green and something yellow all year? Have you investigated indoor growing under lights or in a greenhouse?
- Do you have cool, dry, protected shelves for storage of canned goods away from light?
- Is there a cool (but not freezing) place to put a freezer to cut down on its operation costs?
- Do you own a pressure canner? Have you equipment for open-kettle canning?
- Is there enough sun where you live for outdoor solar drying? Do you have a dehydrator? Could you build one?
- Do you understand each method of preserving?
- Do your growing conditions or your family's food preferences point toward one method of preservation over others?

## Extending the Harvest Season

If you live in the North, you will almost certainly find your late harvests interrupted and eventually terminated by frost. In some areas, a gentle Indian summer period follows the first frost and provides perfect ripening conditions for late crops.

For the gardener, forewarned is forearmed. Check the newspaper and radio for frost-alerts as the date for frost approaches. Look for clear days with high clouds followed by periods of showers, wind shifts, and drops in barometric pressure, followed by a clear day when the temperature doesn't rise above 65°F during the day and drops to 55°F before sunset.

When this sequence of events occurs, gather all ripe and nearly ripe summer vegetables from the garden, including lima beans, peppers, cucumbers, sweet potatoes, and tomatoes. Full-size green tomatoes will go on ripening indoors. Crops that can tolerate some cold weather or even light frost, such as Swiss chard and turnips, can be left in place in the garden as long as they are covered.

Dropping temperatures don't have to put a damper on your gardening endeavors. Just cover vulnerable, tender plants with any of these protectors, and they should come through a cold spell unscathed. From left to right, the items shown here include a bottomless plastic milk jug, a perforated section of a milk carton, a commercial tent-shaped cloche made of wire and glass, a purchased heavy waxed-paper hotcap, and a slatted bushel basket.

Anything that keeps out cold and lets in a little air may be used as a cover in the garden, from upside-down flowerpots to paper bags. Individual plants can be covered with glass jars, upended milk cartons, baskets, or "hats" made of newspaper. An entire row can be protected with a sheet of clear plastic stretched over wire hoops. There is also a variety of glass and plastic covers or "cloches" available commercially to provide frost protection for plants. If there are any plants still growing in your cold frame—a late crop of salad greens, perhaps—replace the cover at night. Be sure to remove all the frost protectors on warm, sunny days.

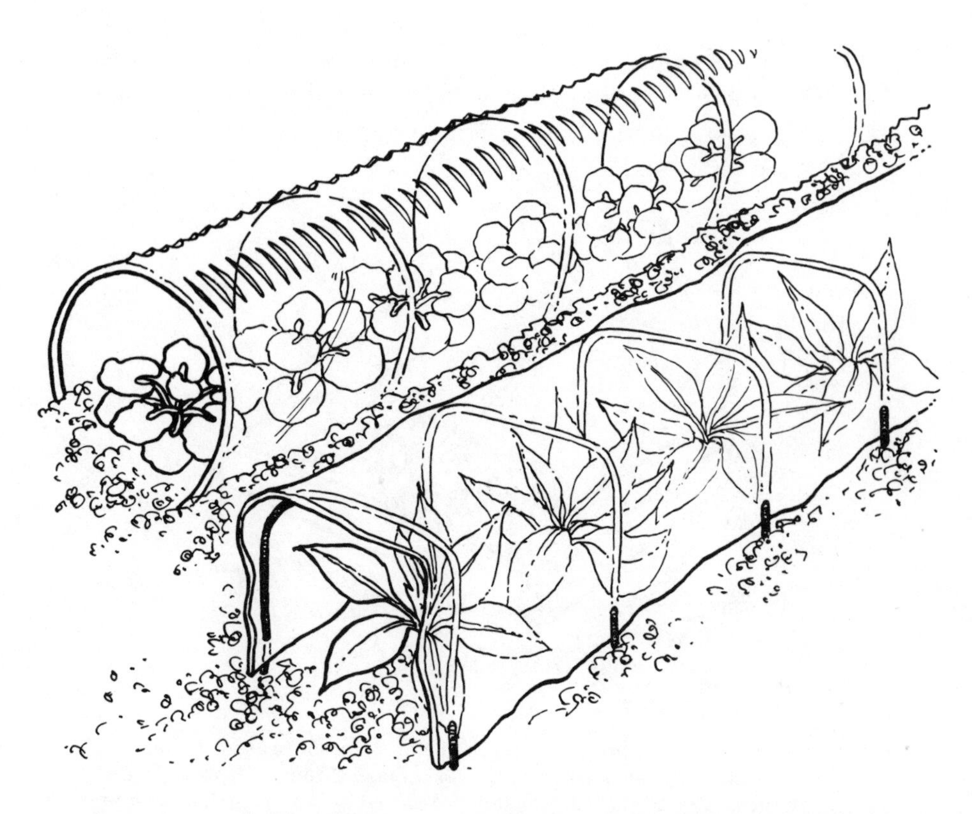

These easy-to-assemble row covers are very effective in providing frost protection. The frame can be made from wire or wood, and the clear polyethylene sheeting can be stapled or tacked in place. Free-hanging sides can be rolled up for ventilation, or the sides can be slitted so that ventilation is automatically supplied.

## Fall Checklist

The precious weeks between the prime harvest season and the first hard freeze are the time to get the garden ready for winter and to begin preparing for the next growing season. These fall weeks are a time of activity in well-managed gardens, and you should take advantage of them. Here are some fall chores to expect:

- Remove crop debris—vines, stalks, frost-nipped plants. Bag any diseased or pest-infested plant parts, and put them out for the trash collector. Shred the healthy ones for compost.

- Till and enrich your garden soil. Fall is the best time to add compost, manure, and rock powders to the garden.

This modest-looking cold frame is a valuable and versatile season extender. The "greenhouse effect," created when light waves pass through glass and turn to heat waves, accounts for the warmer soil and air temperatures inside the frame. So even when it's still too cool to plant outdoors, hardy crops can get off to an early start in the frame. In addition, the cold frame can be used to harden-off transplants and winter over hardy crops.

- Make more compost if the weather is still warm, or stockpile materials in a dry place for the first batch of spring compost.
- Sharpen your tools and give them a light coating of oil for winter protection. Repair any broken handles.
- Bring in soil, sand, compost, and flats to use for starting seedlings in late winter.
- Mulch the garden to keep soil loose and insulate winter-stored crops.
- Grind or shred the leaves you rake from your yard for mulch and compost.

- Check to see that leftover seeds are stored in a cool but not freezing place.
- Remove and store poles, stakes, and trellises.
- Index the vegetables you've stored, frozen, and canned to give yourself a better idea of how fast you use them and how much to plant next year.

## Winter's Song

It's a long, long time from October to March—but not as long as you think. Even in winter, when the snow piles high outside your window, you can be thinking about your garden. Spend a few hours investigating new varieties and vegetables in seed company catalogs. Review the summer's experiences in your mind, or record your thoughts in your gardening journal. Figure out what you learned. Begin work on new lists, plans, and maps. Dream up new ways to stretch the growing season, from starting seeds earlier indoors to investing in a greenhouse. Build or plan projects like indoor shelves for growing plants under fluorescent lights, cold frames, or a root cellar. Lay in compost and mulch supplies. Try out new recipes for stored produce. Save some wood ashes from your fireplace. Think about where to put another garden bed. Learn about insects, birds, exotic plants, advanced gardening techniques. Read books about gardening. Grow plants indoors in pots and tubs. Rest. Dream.

# 5

# How to Grow 20 Favorite Vegetables

This final chapter provides a summary of the cultural needs of 20 commonly grown garden vegetables. You will find specific information on each crop's growing range, soil preference, space needs, and requirements for water and nutrients, along with suggestions for succession planting and companion planting. There are also tips on harvesting and using the vegetables. Use the chapter as a guide to help you in planning your garden and caring for it during the growing season.

## Lima Beans

These beans require three months of warm weather to grow and produce. Gardeners in the far North will have to look for early maturing varieties that will ripen within the shorter season. Like snap beans, limas are available in bush or pole forms. The bush types are very compact and grow to only about 12 inches tall. The pole limas are real climbers, often reaching heights of 6 feet or more. Always be sure to locate these tall growers where they will not throw shade on other sun-loving vegetables in the garden.

Good bush varieties include FORDHOOK and HENDERSON BUSH BABY. Good pole limas are GIANT PODDED, KING OF THE GARDEN, and SUNNYBROOK.

Plant limas in a sunny spot. If you're planting pole varieties, set the poles so they don't shade other crops. Lima beans needs a sandy or very light clay loam enriched with humus that warms up early in spring.

A 100-foot row yields 2 bushels of bush limas or 2½ to 3 bushels of pole limas. To sow a 100-foot row you will need 1 pound of bush variety seeds or ½ pound of pole bean seeds.

Seeds can be stored for three years. Treat seed with a legume inoculant if the planting site has not been used for beans before.

Plant limas only when the soil is thoroughly warm; seeds rot if planted too early. Plant at least two weeks later than early snap beans or in late May in most temperate areas. Plant 1 to 1½ inches deep. Space bush limas 3 inches apart, thinning later to 8 to 10 inches (less in rich soil). Space pole limas 6 inches apart and thin to 12 inches apart. Pole beans grown on teepee supports are thinned to two plants per pole. Always plant eye-side down to strengthen vines and bushes.

Mulch when plants are 4 inches high. Sidedress with compost when 6 inches tall if plants are not mulched. To support pole varieties, use rough-surfaced poles 10 feet long, sunk 1 to 3 feet in the ground. These can be single poles, or groups of three tied at the top to form teepees. A tall fence or trellis may also be used.

Bush limas mature in 65 to 80 days. Pole beans mature in 85 to 95 days. Pick beans as they ripen. Gather them promptly while the pods are still bright green, with seeds discernible through the skin. If beans are to be stored, leave them on the plant to mature and dry. If rainy weather interferes with drying, bring vines laden with beans under shelter to finish drying.

You can use lima beans fresh, or can, freeze, or dry them. Dried beans can be stored for long periods as long as they are kept cool and dry.

## Snap Beans

Snap beans can be had in either compact bush or climbing pole forms. There are myriad varieties, producing green-, yellow-, or even purple-podded beans. Snap beans can be grown wherever there are three frost-free months, although they will not do well in the southeastern or southwestern United States during the height of the hot summers there. Beans are self-pollinating, so two varieties can be grown in the same plot without crossing. Bush beans mature more quickly than pole types, but their harvest period lasts for only a few weeks. Pole beans mature a little later, but you can continue to harvest the beans over a longer period of time.

There are many, many varieties of snap beans. Dependable bush types include BOUNTIFUL, BLACK VALENTINE, IMPROVED TENDERGREEN, TENDERCROP, and TOPCROP (the latter two being mosaic-resistant), and GOLDEN WAX and PENCIL POD WAX (both of which are yellow-podded). Two perennially favorite pole varieties are KENTUCKY WONDER and ROMANO (a flat-podded bean with a rich flavor).

Snap beans need full sun, and do best in a sandy or clay loam that is slightly acid, with high humus content. In alkaline soils watch for zinc deficiencies. Plant pole beans on the north side of the garden so they will not shade other crops. Beans can be rotated with heavy-feeding crops because they help fix nitrogen in the soil.

Yields are approximately 2 bushels per 100 feet of row. For bush beans, 1 pound of seed plants 100 feet of row; for pole beans, ½ pound of seed plants a 100-foot row.

Seeds can be stored up to three years. Germination rate is over 80 percent and occurs in four to seven days when temperatures are between 65° and 85°F. If beans will be planted in soil where peas or beans have never been grown, dust the seeds with a commercial legume inoculant before planting.

Plant most snap beans after the last spring frost date. Sow ½ to 1 inch deep and space bush beans 3 inches apart; space pole beans 18 inches apart in rows along a trellis, or five to six seeds per pole to train on a teepee or A-frame support. Later thin to three or four plants per pole.

It is essential to remove weeds soon after the first bean leaves appear. Hoe shallowly. Mulch when the ground is warm. Never cultivate or work around beans when wet, to avoid spreading mosaic disease.

Plant a second crop when the first planting is above ground, and every two weeks thereafter until midsummer. Follow an early bean crop with a heavy-feeding, late-summer vegetable. Beans make good companions for beets, cabbage, carrots, cucumbers, peas, radishes, or Swiss chard.

Use snap beans fresh, or can, freeze, or dry them for use in casseroles and soups.

## Beets

Beets are heavy-feeding, hardy annuals. In the South they are grown in fall, winter, and spring; elsewhere they can be grown in spring to early summer and in fall.

Beets are one of the least demanding and most rewarding vegetables for the home garden. They are easy to raise from seed sown directly in the garden, and they are not usually bothered by pests and diseases. For the small amount of space they take up, beets offer a bonus harvest—the leaves and early thinnings can be used as greens and are rich in vitamins and minerals.

Early varieties include EGYPTIAN, RED BALL, and RUBY QUEEN. Late varieties include DETROIT DARK RED and WINTER KEEPER. Long-rooted slicing varieties, such as CYLINDRA, are also available, as well as a golden variety.

An open, well-drained, sunny location is best for beets. They need loose, deeply tilled soil that is fertile in the root zone, free of stones, and preferably sandy and humusy. Strongly acid soil is not recommended.

Beets yield about 2 bushels per 100 feet of row. One ounce of seed will plant 50 to 80 feet of row.

Seeds can be stored up to four years. Germination rate is over 60 percent. For better germination, soak seeds for 24 hours before planting.

Plant early beets in March and April, late beets from April to June. Sow early beets ½ to 1 inch deep, late beets 2 inches deep. Space them 4 to 6 inches apart. Beets must be thinned.

Sidedress the plants with compost. Mulch heavily and closely to keep the top 8 inches of soil moist and weed-free. Water after planting to assure germination. Also water plants in dry weather.

Beets can be replanted every three to four weeks until 90 days before the first fall frost. Follow early beets with a late crop of broccoli, cabbage, or head lettuce. Late beets can be planted after an early crop of lettuce or peas.

Beets are good companions for bush beans, members of the cabbage family, lettuce, and onions. Keep them away from pole beans.

Beets mature in 50 to 80 days. Pull them at or before maturity. Beets for winter storage are left in the ground until the first hard frost. Early beets can be pulled when they reach 1 to 1½ inches in diameter. Cut the tops to 1½ inches but leave them attached to prevent bleeding.

Use beets fresh, or can, freeze, or pickle them. Store them with approximately 1 inch of stem attached.

## Broccoli

A member of the cabbage family, broccoli is similar in flavor to cauliflower but is far easier to grow and more nutritious. In most areas it is planted in early spring or in mid to late summer for a fall crop. Summer crops can be grown in northern regions, on seacoasts, and around the Great Lakes. Broccoli needs cool weather to thrive, and, if kept sufficiently moist, will reward the gardener with a large, well-developed central head as well as successive harvests of smaller, tender side shoots.

Many varieties of broccoli are available, including GREEN COMET HYBRID (which matures early), SPARTAN, DECICCO, ITALIAN GREEN SPROUTING, CALABRESE (which is extra hardly), WALTHAM (best for a fall crop), FREEZER, and PROPAGANE.

Broccoli needs cool, open, well-drained land and a constant supply of moisture. It needs moderately rich soil with good calcium content, easy to work and deeply dug. A neutral pH is best.

A 100-foot row will yield 45 heads plus a second cutting of smaller side shoots. A small packet of seed will produce about 200 plants.

Seeds can be stored up to three years. Germination rate is usually 75 percent or better.

Broccoli can stand a light freeze, so 5-inch seedlings can be planted outdoors as soon as soil is worked in spring. Start seeds indoors in midwinter. For a fall crop, start seeds in early summer (late May through early June). Seeds should be sown ¼ inch deep; set seedlings slightly deeper than they were in the flat or pot, about 3 inches deep. Early broccoli needs 1½ feet between plants and 2 feet between rows to prevent shading. Late plantings should be spaced 2 to 2½ feet apart in rows with 3 feet between them. Homegrown seedlings benefit from transplanting before they are ready to set out.

Broccoli is a heavy feeder with a wide, deep root system. When transplanting, dig holes big enough for roots and place abundant compost in the root zone. Topdress with compost when setting out plants, and again when heads begin to form on fall crops. Shield newly planted seedlings from hot sun for the first week. Plants benefit from heavy mulch. Water frequently in dry weather and when heads begin to form. Soak thoroughly when setting out new plants.

Never follow broccoli with another cabbage family member. Intercrop with carrots, lettuce, onions, spinach, or Swiss chard. Plant broccoli after early peas.

Broccoli matures in 40 to 70 days. Be sure to cut heads before any yellow flowers appear (the head is actually made up of tiny green flower buds). After the main head is cut, a second crop of small florets should appear in the left axils. To get a second harvest, leave the bottom stalk after cutting at least 4 inches of stem along with the main head. Four to six cuttings are possible from a single plant. Soak heads in salt water before cooking to drive out any tiny green cabbage worms that may be hiding inside.

Use broccoli fresh, either cooked or raw in salads, or freeze it.

## Cabbage

Cabbage grows in all parts of the United States and much of Canada, with fall and winter plantings only in the South. It is generally frost-hardy to 20°F. Cabbage is easy to grow as long as its requirements for cool temperatures, plenty of moisture, and lots of nutrients are met. Many gardeners don't realize that a second harvest can be coaxed from the stumps of early varieties that have already been harvested. Tiny sprouts will appear on the cut stump. Just rub out all but one of these and let it develop into a small but tasty head.

There are lots of good cabbage varieties available, including EARLY JERSEY WAKEFIELD, GOLDEN ACRE, and MARION MARKET (all early varieties), late-maturing DANISH ROUNDHEAD, SAVOY KING, and RUBY BALL and MAMMOTH RED ROCK (both red varieties).

Plant cabbage in a cool, moist, well-drained spot. Cabbage needs full sun to mature. It grows best in humus-rich, moist, crumbly loam soil that drains well.

Early cabbage produces about 100 pounds per 100-foot row; late types yield 175 pounds in the same space. A quarter-ounce of seed will plant a 100-foot row. A 100-foot row will accommodate 70 plants of early-bearing cabbage or 60 plants of late types.

Seeds can be stored up to five years. Germination rates average over 75 percent.

Plants of early cabbage should be set out in spring (April to mid-May, depending on your location). Sow seeds indoors in February or March if growing your own seedlings. Set out late cabbage no later than August 1. To plant, dig holes wide enough to accommodate seedling roots and deep enough so you can put a layer of compost in the bottom. Water well and firm the soil when planting. Protect late transplants from the hot sun for the first week.

Cabbage is a heavy feeder that benefits from sidedressings of compost and from manure spread in fall. Water during dry weather, and mulch to keep the soil moist. To help prevent cracking of heads, make sure the mulch doesn't remain wet right near the plants. When the head is nearly mature, break feeder roots near the surface by cutting them with a shovel.

Late cabbage may be planted to follow early peas or carrots. Early cabbage may be followed in the garden by beans, beets, or late corn.

Set cabbage in or between rows of beets, spinach, lettuce, or other small salad crops.

Cabbage matures in 60 to 120 days, depending on the variety. Cut heads of early cabbage as needed. Harvest late cabbage after the first frost.

Early cabbage is best when used promptly. Late cabbage can be stored in straw-lined trenches or outbuildings, roots up, covered with a layer of straw. Cabbage can also be quartered, shredded, and pickled for sauerkraut, or frozen or canned.

## Carrots

Carrots are grown throughout the continental United States and in Canada. They can be planted from spring until midsummer in northern areas and are grown as a fall and winter crop in the South. They do best when air temperatures is in the 60s.

Tough, pithy supermarket carrots cannot compare to the homegrown kind. When you grow your own, you can harvest them while they are still young, tender, and at their sweetest. If the garden soil is not deep and crumbly, try growing the shorter, stubby-rooted or round-rooted varieties, which don't need as much depth to develop well.

A huge selection of carrot varieties is available, in a wide range of shapes and sizes. Try DANVERS, GOLDINHEART, and NANTES (medium- to large-rooted), LITTLE FINGER and OXHEART (small), CHANTENAY, IMPERATOR, and GOLD PACK.

Carrots need an open, sunny location with deep, well-drained, loose soil. A rich, sandy loam, enriched with humus and deeply dug, is best. Make sure the soil is free of stones and clods. Early carrots need richer soil than later-maturing crops. A slightly acid pH is preferred.

The tiny seeds are difficult to handle, and you will need from ½ to 1 ounce of seed to plant a 100-foot row. A row of that length will give you about 2 bushels of carrots.

Seeds can be stored up to three years. Germination rate is in excess of 50 percent. For faster germination, soak seeds and pregerminate them between layers of moist paper towels and plant when the white root tip first appears. Some gardeners mix the tiny seeds with sand for easier sowing.

In most areas, plant carrots from early spring until midsummer. In the South plant them in fall. Sow early crops ½ inch deep; crops planted later should be seeded 1 inch deep. Cover newly planted seeds with sifted compost. Space seeds ½ inch apart. Because the seeds are so difficult to handle, you will probably have to thin the seedlings to this distance when they are 2 to 3 inches tall. Thin again when roots are ½ inch in diameter to stand 1½ to 2 inches apart.

When the plants are several inches tall, mulch them heavily, especially fall and winter crops. Cover the shoulders of young carrots with soil when you thin for the second time. Water thoroughly during dry spells.

Plant seeds every three weeks from early spring until 2½ months before the first fall frost. Early carrots may be succeeded by tomatoes or peppers.

Carrots make good companions for bush beans, lettuce, onions, peppers, radishes, or staked tomatoes.

Carrots mature in 70 to 75 days, but they can be pulled when smaller. You can dig or pull carrots anytime after the roots are ½ to 1 inch in diameter. For winter storage, dig carrots after the first frost, when the ground is dry. Remove the tops and store the roots upright in a box filled with sand. Do not store any damaged carrots, and do not let those in the box touch one another.

You can use carrots fresh, or freeze, can, or dry them in slices.

## Corn

Corn is a warm-weather crop that matures fastest when temperatures are up around 90°F. It can be grown anywhere except in far northern areas.

While there's nothing tastier than a freshly picked ear of corn, many gardeners find their enjoyment usurped by hungry raccoons, who seem to sense the moment when the corn is ripe and ready. To thwart these unwanted diners, try growing vining crops like squash and pumpkins around the corn. All the tangled, viny growth makes the raccoons uncomfortable since it cuts down on their visibility and slows down their getaway. This should be enough to deter them so that you, instead of the local wildlife, get to enjoy the harvest.

Good corn varieties include EARLY SUNGLOW, GOLDEN BEAUTY, and NORTH STAR (early yellow types), HONEY AND CREAM and PEARLS AND GOLD (which have both white and yellow kernels), GOLDEN CROSS BANTAM, GOLD CUP, and HONEYCROSS (midseason yellow varieties), DELICIOUS and EVERGREEN (both late varieties), HYBRID COUNTRY GENTLEMEN and SILVER QUEEN (white kinds), and COUNTRY GENTLEMAN and GOLDEN BANTAM (which are open-pollinated varieties from which you can save seed). In addition, there are also the newer extra-sweet varieties.

Corn needs a sunny, well-drained site. Hillsides are good, and so is newly dug land. The best soil for corn is a deep, mellow, humus-enriched, neutral to slightly acid loam. Calcium, in the form of powdered limestone, should be added if the soil is deficient. Potassium supplements of wood ashes or granite dust may also be required.

A 100-foot row of corn (about a pint of seed) will produce 50 to 60 ears.

Seeds can be stored for up to three years. Germination rate is generally better than 75 percent.

Plant early varieties just after the last frost date, unless you will be planting in a wet clay soil, which will delay germination. In that case, wait a few weeks to let the soil warm a bit before planting. Plant late crops 80 days before the first fall frost if you are planting a variety that matures in 65 days. The main crop does best if planted when soil temperature is around 62°F.

To enhance pollination, plant in blocks two or three rows wide. Space seeds 3 to 4 inches apart in rows when sowing; later thin to 12 inches apart for tall varieties, 8 inches apart for shorter-growing types. In organically rich soil, closer spacing is possible. Plant in hills, six seeds to a hill, later thinning to three plants per hill. Hills should be 18 inches apart. Push soil close around maturing stalks as the plants grow.

If you are planting in hot, dry weather, mulch the ground before planting to assure moisture for germination. When planting, pull back the mulch and replace only a 1-inch-deep layer after seeds are planted. When plants are 8 inches tall, the mulch can be 6 inches deep. If you do not use mulch, rake the soil surface three days after planting and hoe away the weeds, being careful not to disturb the roots.

Put compost in the bottoms of planting furrows or hills. When stalks are knee-high, sidedress with compost or well-rotted manure.

Corn needs plenty of water from ten days before tasseling through the flowering period. After that, withhold water.

For a continuous crop, plant an early variety every two weeks, or plant several varieties with different maturing times at once. Never follow corn with a heavy-feeding crop. Rotate to beans or, if the same garden area is to be used for corn repeatedly, use an alfalfa cover crop in fall to be tilled under before the next planting.

Carrots, head lettuce, onions, or other early spring crops that can stand shade may be interplanted with corn.

Corn matures in 65 to 100 days, depending on the variety. Pull back the husk and check the kernels. If milk spurts when a kernel is pressed with your fingernail, the ear is ripe. Corn loses its sweetness quickly after picking, so pick it right when you are ready to use it. Twist the ears from the stalks.

Use corn fresh, or freeze it on or off the cob, or can it.

## Cucumbers

Cucumbers can be grown throughout the United States, but gardeners in the South would do well to grow them only in cool weather. These crisp, cool, and refreshing vegetables are 95 percent water. Where garden space is tight, grow compact bush varieties or train vining varieties to grow up a trellis or fence. A fenceful of cucumbers can form an attractive and effective screen to shade a patio or hide a woodpile or compost bin from view.

Cucumber varieties abound, and this listing is but a fraction of those offered. You can choose from BURPEE HYBRID and GEMINI HYBRID (both disease-resistant), pickling types like WISCONSIN, SMR-18, and MARKETER, POINSETT (which is good for southern gardens), and CHINA LONG (an oriental type that grows up to 2 feet long). Two varieties for container growing are PATIO PIC and POT LUCK.

Cucumbers grow best in moist but well-drained locations. They tolerate partial shade. A warm, sandy, humus-enriched neutral or slightly acid loam is ideal.

A 100-foot row will yield about 1½ bushels of cukes. A half-ounce of seed is adequate for 100 feet of row.

Seeds can be stored up to five years. The germination rate is 80 percent or more.

Plant seeds in the garden when all danger of frost is past. Cucumbers can be started earlier indoors, but the seeds require bottom heat to sprout and the seedlings transplant poorly. Sow seeds ½ inch deep. Space bush varieties 6 to 12 inches apart, and vining varieties 20 inches apart. In flat-topped hills, grow three plants to a hill, with 5 feet between hills. In mounded hills, plant six to nine seeds in each 2-foot-diameter hill, thinning to four plants when they reach 4 feet.

Cucumbers are heavy drinkers. Water well at least weekly during drought. Apply a deep mulch. If wet soil deters early mulching, cultivate with a hoe when the vines are 3 inches long and keep the ground loose until they reach 18 inches. By that time the soil should be dry enough to mulch. Provide constant nourishment for these heavy feeders by planting in compost-filled holes or trenches, and use compost or manure in preparing the plot the fall before planting. Otherwise, sidedress with compost when the vines are 1 foot high. If the planting is early or the harvesting late, provide frost protection in the form of cloth or plastic covers or heavy, temporary mulch. A sturdy fence, trellis, or net support will keep fruit off the ground where it could rot.

Good companions for cucumbers include beans, members of the cabbage family, peas, radishes, and tomatoes.

Cucumbers mature in 55 to 75 days. Watch developing cukes closely; they swell so fast you can almost see them growing. Pick the first small fruit to encourage better production. For pickles, pick when fruits are 2 to 6 inches long; for table use, pick when 6 to 10 inches long. Oriental varieties grow longer—as much as 2 feet. Pick them while they are still deep green in color.

Slice cucumbers fresh for table use, or process them into pickles and relishes. Cucumbers do not freeze well.

## Lettuce

This salad mainstay grows throughout the United States and in Canada, but only when temperatures remain cool. In the South lettuce is an early spring and winter crop; in the North it is a fall and spring crop, although some looseleaf strains will keep growing through the summer.

Lettuce grows best in a lightly shaded spot, so it is the ideal crop to tuck into those places in the garden that don't receive full sun. Lettuce, especially the looseleaf types, is a good succession crop because it grows quickly and can precede and often follow midseason crops.

Lettuce comes in looseleaf and heading varieties. Good looseleaf varieties include BLACK-SEEDED SIMPSON, GRAND RAPIDS, SALAD BOWL, RED SAILS, and OAK LEAF. Good butterhead types are BIBB, BIG BOSTON, and DARK GREEN BOSTON. A heat-resistant cabbagehead variety to try is PREMIER GREAT LAKES (also widely known as ICEBERG). Cos or Romaine has tall, conical heads; PARIS WHITE is one reliable variety.

Lettuce prefers a cool, well-drained location that will remain moist. A rich soil, dug to 8 inches and enriched with compost, promotes best growth. Early varieties prefer sandy loam, later ones like clay loam. A neutral to slightly acid pH is best.

A 100-foot row will produce about 80 heads or 50 pounds of leaf lettuce. One small packet of seed will plant the row.

Lettuce seed can be stored for up to six years. Germination rate is 80 percent or better. When the garden soil is hot, presprout the seeds in the refrigerator on wet paper towels.

Seed can be sown very early, in March or April in most areas. Fall crops can be planted in August and September. For an extra-early spring crop, plant in late September and mulch heavily. Sow seeds ¼ inch deep. Lettuce needs light to germinate, so some gardeners broadcast the seed and just rake lightly to cover it. It's almost impossible to sow the fine seeds thinly enough, so try planting in pinches of seed 6 inches apart. Thin when the plants have four leaves and are big enough to handle. Thin looseleaf varieties to 6 inches apart. Thin heading types to stand 10 to 16 inches apart. Lettuce is easy to start indoors and transplant outside. Heading types require 80 to 90 days of cool growing weather, so they often need an indoor start to beat the hot days. Plant in flats kept at 50° to 60°F. Thin when large enough to handle, and make sure no two plants touch. Transplant and thin again when plants are 2 inches tall, and plant them out when they reach 6 inches.

Lettuce needs almost as much water as cucumbers. Water regularly during dry weather, but be sure you do not wash out roots with too hard a spray. Keep down weeds with mulch, or hoe shallowly and with care. Rake some humus into the row or bed before planting, and sidedress with sifted compost. Seedlings planted in late spring need shade through their heading period since they will only head when kept cool. All transplants need shading temporarily after being set out.

Plant lettuce every week or two until the end of May, to provide a long harvest. Follow with any midseason crop, such as corn, peppers, or tomatoes.

Good companions for lettuce include other small salad crops, beets, members of the cabbage family, and onions.

Lettuce matures anytime from 45 to 85 days. Looseleaf kinds can be picked many times if only the lower leaves are taken. Pinch off the top center to keep the plants from bolting. Pick head lettuce in the early morning to assure freshness.

Lettuce is for fresh use only; it does not can or freeze well.

## Onions

These flavorful vegetables can be grown throughout the United States and in Canada. They need cool weather to develop foliage, so southern and western gardeners should plant them in fall or winter. Northern gardeners can start seeds indoors to allow time for the bulbs to develop during long days and hot weather, or they can start with bulb sets or purchased seedlings. There are two types of onions: bunching onions, which are grown for use as scallions, and storage onions, which are grown for their large bulbs. Storage onions pulled early can also be used as scallions.

To start from seed, dependable bulb-forming varieties include SWEET SPANISH, EARLY YELLOW GLOBE, YELLOW GLOBE DANVERS, and YELLOW BERMUDA. To start from sets, good bulb-forming varieties include EBENEZER and ITALIAN RED. Two good bunching onions are EVERGREEN LONG WHITE BUNCHING and BELTSVILLE BUNCHING.

Grow onions in a well-drained, moist area. They need a crumbly, moderately acid, sandy soil that has plenty of nutrients in its top layer. Silty loam or peat soil is good, too.

Yield is 1½ bushels (roughly 80 pounds) per 100-foot row. For a 100-foot row, count on using ½ ounce of seed, 2 pounds of sets, or 600 seedlings.

Stored onion seed stays good for only a year or less. Germination rate is around 70 percent.

Sow seeds very early in spring, even before the last frost. Seed germinates best at 65°F. For earliest green onions, about a week after the last frost, set out seedlings started inside or purchased. Late onions can be sown in mid-June in the North. Sow seeds ½ to 1 inch deep. Plant sets 1 to 2 inches deep. Set seedlings slightly deeper than they were in flats. Space seeds, sets, and seedlings 2 to 3 inches apart, or closer in rich soil. Thin large onions so they don't crowd each other. Plant seed thickly, and cover with sifted compost. Water well. Sets should be covered with an inch of compost, with the tips just under the soil surface. Reject any sets larger than a dime—they are apt to send up flower stalks. Start plants from seed indoors ten weeks before setting out. Plant when they are ⅜ inch in diameter.

Water onions only during dry spells. Cultivate shallowly. When the plants are 10 to 12 inches tall, draw mulch close to the plants to keep the bulbs from splitting. Use compost in trenches or holes when planting. Sidedress each month.

You can use onions to fill spaces in the garden between widely spaced vegetables like beets, broccoli, late cabbage, peppers, Swiss chard, and tomatoes.

Bulbing onions take 120 days to mature from seed, 100 days from sets. Bunching onions are ready in 55 days. Pull bunching onions as needed when they are at least 8 inches tall. As storage bulbs mature, the tops begin to curl and wither. When most of the leaves have already fallen, help the others along by stepping on them. A week later, pull or dig the onions and cure them on top of the ground if the weather is fair (indoors if it's rainy).

When the bulbs are thoroughly dry, move them to a cool, dry cellar and store them in ventilated containers. Onions may be touched lightly by frost without being hurt, but they should be used soon after freezing. Don't handle them when frozen.

The best storage method is to keep onions in their cured state. A decorative way to keep them is to braid their tops (add string for extra strength) and hang them in the kitchen. The best keepers—and the best for braiding—are late-maturing onions with thin necks. Two less satisfactory storage methods are freezing and canning. Dice before freezing.

## Peas

Peas are cool-weather plants that are traditionally among the earliest crops planted in the North; they are an early spring and fall crop. In the South and Southwest, peas are planted in the fall. Both dwarf and vining plants are available. Shell peas are grown for the peas inside—the only edible part. Edible-podded peas, such as snow peas, are just that—they can be eaten pods and all, before the peas fully develop inside. Sugar snap peas have a crunchy, delicious pod, but can also be eaten as shell peas.

Dwarf shell pea varieties include LAXTON'S PROGRESS and LITTLE MARVEL. Tall varieties include ALASKA, ALDERMAN, FREEZONIAN, LINCOLN, and THOMAS LAXTON. Edible-podded types include MELTING SUGAR, and SUGAR SNAP and its variants.

Peas can grow in shade, but should not be grown in the same location more often than once every four years. The best soil for peas has plenty of humus, potassium, and phosphorus, and is neutral or slightly acid. Early crops prefer soil with more clay.

A 100-foot row produces a bushel of shell peas or 2 bushels of edible-podded peas. One packet of seed will plant about 20 feet; 1 pound will plant about 100 feet.

Seeds can be stored for three years. Germination rate is 80 percent or better. Soak seed briefly when planting weather is dry, and water thoroughly when seeding. Treat with legume inoculant before planting if peas or beans have not previously grown on the land.

Sow early varieties as soon as the ground can be worked in spring, or up to six weeks before the last frost. Sow later peas every two weeks until mid-May. Fall plantings may be made in late July to mid-August, depending on the date of expected first fall frost. Cover with 1 to 2 inches of sifted compost. The lighter the soil, the deeper the planting. Space seeds 1 to 4 inches apart. Peas may be planted in double or quadruple rows 6 to 8 inches apart, leaving room for mulch. Thin them to 3 or 4 inches apart.

Mulch deeply, and water from the time blossoms fall until the peas are picked. In lieu of mulch, weed frequently or cultivate shallowly. Dwarf varieties may be propped up with thick mulch so the vines won't rest on wet ground. Double rows of dwarf plants will support each other. All varieties, especially tall ones, benefit when given additional support. Use twigs, set in place when seeds are planted, for low vines. For taller varieties, chicken wire fences supported by stakes work well, as do trellises or commercial plastic support netting. Secure the vines to the support at 1-foot intervals.

Peas can be followed by corn, late cabbage, or other late-maturing crops.

Good companions for dwarf peas are beans, carrots, cucumbers, green onions, lettuce, radishes, or spinach.

Peas are ready to pick in 65 to 80 days. Never handle plants when they're wet. Pick shell peas when the pods are well-filled but peas are still small and tender. Edible pods are picked when the pods are fully grown but still flat. Sugar snaps can be picked any time from when they are a few inches long until the pods are plump with peas. Refrigerate peas immediately after picking—they lose their sweetness quickly. Wash before—not after—shelling.

Use peas fresh, frozen, canned, or dried in soups. Process quickly to preserve vitamins.

## Peppers

Peppers can be perennial in the Deep South, but they are frost-tender, so far-northerners can't grow them unless they are started indoors from seed. Peppers need temperatures of 65° to 80°F to set fruit. Both hot and sweet peppers can be grown in the same way.

Sweet pepper varieties turn from green to red or yellow as they ripen; they can be picked at either stage. Red peppers contain more vitamin A and tend to have a sweeter flavor. However, if you leave the peppers on the plant until they turn red, the plant will produce fewer of them.

Reliable sweet varieties include BELL BOY HYBRID (which is mosaic-resistant), CALIFORNIA WONDER, KING OF THE NORTH, PIMENTO, and YOLO WONDER. Hot varieties include CAYENNE, HUNGARIAN WAX, LONG RED, and RED CHILI.

Peppers need a very sunny, well-drained, open spot. They prefer organically enriched, loose soil, not overly high in nitrogen. Soil should have a layer of good organic material on top of a gravel subsoil.

A 100-foot row yields 4 bushels. A small packet of seed is adequate for the row.

Stored seed is good for two years. Germination rate is around 55 percent or better. Pepper seed is difficult to germinate, and needs heat and moisture for best results.

Start seeds indoors five to eight weeks before the last frost. Set plants out when all danger of frost has passed, or two weeks after the last frost date. Sow seed ½ inch deep. Set seedlings in the garden slightly deeper than they were in flats, or up to within 1 inch of the first leaves. Space seedlings 24 inches apart; closer in rich soil.

Water peppers heavily when young; lightly for the rest of the growing season. Mulch or cultivate shallowly when young.

Peppers can follow early lettuce or beets in the garden. Carrots, onions, and tomatoes are all good companions.

A mature pepper is heavy and firm to the touch. If you want to harvest green peppers, pick them just before they turn red. Cut stems with a sharp knife, ½ inch from the cap.

Peppers are best fresh, and green ones can be stored up to 40 days in ventilated baskets at 32°F under very humid conditions. Peppers may be frozen without blanching, canned, or dried. Dry chili peppers by hanging the vines upside down in a well-ventilated place. Tie the pods on strings and hang them in the kitchen.

## Potatoes

Potatoes are usually grown only in the North, but early crops of some varieties will grow in the South, at least to new potato size, if they're mulched.

You may be tempted to take a few potatoes you bought at the market and cut them up for planting. But most potatoes sold in the store have been chemically treated to inhibit sprouting, and would only rot in the garden. For best results, buy special seed potatoes for planting. If you save potatoes from one year's crop to use for next year's, there is the risk that they harbor virus diseases which will reduce future harvests. It's best to start fresh with new seed potatoes every year.

Make a point to select disease-resistant varieties. Good early potatoes include IRISH COBBLER, NEW NORLAND, and NORGOLD RUSSET. Good late varieties for storage are BUTTE, GREEN MOUNTAIN, KATAHDIN, KENNEBEC, and SEBAGO.

Grow potatoes in a well-drained, moist location. Never follow tomatoes or use recently limed soil. The best soil is a fertile, moderately acid, sandy loam.

A 100-foot row will give you about 3 bushels of potatoes. You will need 7 pounds of seed potatoes to plant a 100-foot row.

Store seed potatoes over the winter only. Germination rate is 95 percent in well-drained areas. Spread the seed potatoes in a warm shed or on a porch or other location that gets sunlight and where it is about 60°F. Turn them occasionally until they are green on all sides. Sprouts of ¼ inch will develop from the eyes. Discard potatoes with spindly sprouts. Either plant small whole potatoes, or cut larger ones into egg-sized pieces, each containing one or two eyes. Be sure to cut ample tissue with the eyes, and harden the potato sections before planting them.

Plant an early crop as soon as the ground can be worked in spring, five to six weeks before the last frost. Plant late crops early enough so they mature three to four weeks before the first fall frost. Plant 1 to 3 inches deep, depending on the lightness of the soil; cover them deeper later on. Some gardeners successfully plant potatoes in a bed of mulch with more mulch heaped 10 to 12 inches over them. If you plant too close to the surface, the crop potatoes will turn green from the sun and become toxic. Plant seed potatoes 12 to 15 inches apart.

If the weather is dry, water occasionally until the plants flower. When the tops begin to yellow, stop watering. When shoots are 9 inches tall, draw soil around the plants, leaving no more than 6 inches of stem exposed above ground. This is done to prevent all light from reaching the tubers, which will form above the seed eye. Increase the hill size, making a flat mound every three weeks until flowering, or use a deep mulch for the same purpose. If your soil requires it, be sure to add potassium to the soil in the form of greensand or granite dust before tilling. Lack of potassium can cause mealy, soggy-tasting potatoes.

Early crops of potatoes can be followed with a second late crop, or with late turnips. Early sweet corn or late cabbage can be planted between rows of early potatoes. Beans, peas, and horseradish also make good companions.

Potatoes mature in 100 to 120 days. Gather early potatoes for use in the kitchen as needed after blossoms have formed. Reach into the soil or mulch for new potatoes, and leave the plant undisturbed. Later, harvest the rest of the crop by digging the whole plant. Leave late storage potatoes in the ground until the vines have died down. Dig them with a spading fork before frost or soon after, when the ground is dry. Dry the tubers on top of the ground for an hour to free them of mud. Put the potatoes immediately into a dark, well-ventilated place. In very dry weather, you can omit the outdoor drying period. If the weather turns warm and wet anytime after vines have withered, dig the potatoes to prevent sprouting.

Store late potatoes at 36° to 40°F in a dark, humid but well-ventilated place. Some potatoes will keep for a full winter and into early spring before they sprout and become soft.

## Sweet Potatoes

Sweet potatoes need warm nights to mature; therefore, they are seldom grown commercially north of New Jersey. Home gardeners in northern areas can try their hands at a crop by choosing early-maturing varieties especially suited for colder climates.

Sweet potatoes belie the adage that anything that tastes good must be bad for you. These sweet, moist, and tender tubers provide plenty of vitamins A and C, plus lesser amounts of protein, calcium, and iron. Gardeners who are cramped for space can still afford to grow this nutritious vegetable by selecting a bushy variety rather than a standard vining one.

Varieties include ALL GOLD, APACHE, CENTENNIAL, and NANCY HALL (both of which are good for southern gardens), NEMAGOLD (resistant to nematodes and suited for northern gardens), ORLIS, PORTO RICO, and RED GOLD.

Sweet potatoes grow well in airy, open, sunny sites that hold warmth. The best soil is a warm, sandy, loose loam that is deeply worked. Clay subsoils are best, and the soil should be rich in phosphorus and potassium.

A 100-foot row will produce about 2 bushels of potatoes; 100 slips (rooted shoots) plant a 100-foot row.

Start sweet potatoes from slips, sprouted indoors or purchased. At home, half-submerge plump, moderate-size sweet potatoes in a pan of water and keep them on a brightly lit windowsill at 75°F for a month before planting time. Twist slips from the parent potato when they reach 6 to 9 inches long and have well-developed leaves and root clusters.

Plant slips when soil is thoroughly warm, at least ten days after the last frost date. Vines may die if subjected to temperatures below 50°F. Set purchased slips lower than they grew in pots. Set homegrown slips 4 to 6 inches deep. Space slips 1 foot apart with 3½ feet between rows. For easy harvesting, plant on slightly elevated ridges 6 to 10 inches higher than the surrounding soil.

Water heavily after slips are transplanted to the garden, and again when vines begin to run, then cut back on watering. Mulch to block out weeds, or hoe shallowly with care. The plants will eventually shade out weeds themselves. Sidedress when plants run.

Sweet potatoes need 175 frost-free days. In northern gardens they sometimes fail to mature, but in other years they will bear an adequate crop before frost. Dig the tubers with a fork just before the first frost is expected. If the plants get caught by frost, trim off blackened leaves and cut back to the ground if badly damaged. The potatoes will turn bitter if the plants are not trimmed. Dig the potatoes as soon as possible and cure them in the sun for several hours. Put them in a well-ventilated place with 90°F temperature and very high humidity and leave them there for 15 days. When the curing is finished, store the tubers at 50°F in a place with high humidity. Handle the sweet potatoes as little as possible. They should keep well for several months.

Sweet potatoes can be used fresh, or frozen or canned after cooking.

## Radishes

This root crop has the crispest texture and snappiest flavor when grown in cool temperatures. Excessive warmth causes radishes to turn hot in taste and pithy. Summer and winter varieties can stand more heat than the spring types. Southern gardeners grow radishes in winter, fall, and early spring, and harvest before late spring, or plant in midsummer for a fall harvest. Spring radishes are the well known red balls; summer and winter varieties are long and tapered and can be pink, white, or black in color.

Spring radish varieties include CHAMPION, CHERRY BELLE, FRENCH BREAKFAST, and RED BOY. Among the varieties of summer radishes are SCARLET KING and WHITE ICICLE. Winter radishes include BLACK LUXURY, BLACK SPANISH, and WHITE CHINESE.

Grow radishes in an open, well-drained spot. A light, near-neutral pH sandy loam with lots of nutrients in the upper layers is best. Clear away all stones and enrich with humus. Rich clay loam is best for late varieties.

A 100-foot row yields about 100 bunches of radishes. Use ½ ounce of seed to plant a 100-foot row.

Radish seed can be stored for four years. Germination rate is 75 percent or better.

Plant early varieties around April 1 in temperate areas. Make small, frequent plantings every 10 days until the weather becomes hot. Start planting late varieties in late May or June and repeat plantings until mid-August. Make ½-inch-deep rows and fill with 1 inch of compost, sow the seed, and cover with soil. Sow late varieties ¾ inch deep. Allow ¾ to 1 inch between early red varieties for shoulder-to-shoulder spacing. Larger late varieties should be seeded or thinned to stand 4 to 6 inches apart.

Give radishes a steady supply of water to provide adequate moisture content in the roots to keep the vegetable crisp and mild, and to keep the plants from bolting. Mulch to keep the soil cool. Sidedress early varieties with compost in addition to using sifted compost in the bottom of the row.

Plant every week to ten days in small quantities to get a

continuous crop, or follow early plantings with any late-season crop. Radishes are often used to mark rows where later-germinating crops are planted. The radishes are harvested before the later crops need the additional growing space. Plant radishes with any small salad crop, or with parsnips or beans, carrots, peas, or herbs.

Spring radishes mature quickly, in 20 to 30 days. Summer and winter radishes take 30 to 60 days to mature. Pull promptly when the roots are of usable size.

Use small radishes fresh. Store winter radishes in a cool (32° to 40°F) place with very high humidity.

## Spinach

This cool-season leafy green is grown as a very early spring or late fall crop in the North. In the South, it is planted in late winter or in the fall and wintered over under mulch.

Spinach must be planted so that it matures in cool weather. After several days of warm temperatures, the plant will begin to produce flower stalks, which means that the flavor of the leaves is declining. Careful planning of planting dates is worth the effort, since spinach leaves are rich in vitamins A and C, as well as a host of minerals.

LONG STANDING BLOOMSDALE and VIRGINIA SAVOY (with crinkled leaves), KING OF DENMARK and NOBEL (smooth-leaved), and HYBRID NO. 7 and WINTER BLOOMSDALE (disease-resistant) are some recommended varieties.

Spinach grows best in an open, well-drained location. It tolerates partial shade. A heavily enriched, sandy, well-drained soil promotes good growth. Slightly alkaline soil is preferred.

A 100-foot row of spinach yields about 40 pounds of greens. One-half to 1 ounce of seed will plant the row.

Unused seed can be stored for up to three years. Germination rate is 60 percent or more. If the ground at planting time is dry and hot (especially for fall planting), soak seeds and presprout them between damp paper towels for about a week. Mix the seeds with dry sand before you plant them.

Plant spinach as early as possible in spring, even if you must dig furrows in partly frozen earth. A February thaw may provide an opportunity for extra-early planting. The seed is hardy but will not germinate in heat, and the plants will bolt even in moderately hot weather. The earlier you plant spinach, the better your crop will be. For a continuous crop, plant every ten days until mid-April in the North. In fall, sow from mid-September until the ground freezes. Mulch the last planting heavily for early spring production. Sow seeds ½ to ¾ inch deep, allowing 2 to 4 inches between plants. Spinach is often planted in double rows 8 inches apart with 4- to 6-inch spacing within rows for mature plants.

In dry weather, water every three to four days. Mulch with grass clippings for added nitrogen after a few leaves have started growing.

Spring crops of spinach can be followed in the garden by late beans, corn, or tomatoes.

Good companions for spinach include beans, members of the cabbage family, or other small salad crops.

Spinach is ready to pick in 40 to 70 days. Usually the whole plant is cut or pulled, but if only the outside leaves are harvested, the plant will continue to grow leaves and the harvest can be extended. If you opt for this method, be careful when picking individual leaves to not harm the plant. Plants are considered mature when at least six of the leaves are 7 inches long.

Use spinach fresh, frozen, or canned. Wash fresh spinach quickly and cook it without additional water (the water that clings to the leaves is enough), until the leaves are just tender.

## Squash

The warmth-loving squashes grow wherever there is a long growing season. Squash is not feasible in the far North unless started indoors from seed. Summer squash is grown for the immature fruits that develop relatively quickly and keep on coming until season's end. Winter squash takes longer to mature, because it must develop the hard skins that make it such a good vegetable to store. Plant more winter than summer varieties, since they store so well for so long. Space-saving bush types are available in both summer and winter varieties, in addition to the standard vining types.

There is a seemingly endless number of summer and winter squash varieties available. Dependable summer squash include WHITE BUSH SCALLOP, SCALLOPINI, GOLDNECK, EARLY PROLIFIC, STRAIGHTNECK, BLACK ZUCCHINI, and COCOZELLE BUSH. Good winter squash are BLUE HUBBARD, BUTTERCUP (a turban-shaped variety), TABLE QUEEN (what we know as acorn) and BUSH TABLE QUEEN.

Squash needs a well-drained, sunny spot. It will grow among corn, in grass or weeds at the garden's edge or, if the fruit is supported, on a fence or trellis. A light, sandy loam that's rich in humus and slightly alkaline is preferred.

From a 100-foot row you can expect about 135 pounds of summer squash or 30 pounds of winter squash. You will need ½ to 1 ounce of seed to plant the row.

Unused seed can be stored for four years. Squash germinates at a rate of 75 percent or more.

Plant summer squash when all danger of frost is past through early summer, or in April and May if the ground is not cool and damp. Winter squash is planted in May and June in most regions. Sow seeds ½ to 1 inch deep. Some mature vines need 6 feet on all sides; most will need about 3 feet. Bush types can get by with just 1 foot on all sides. Squash can be planted in hills, six seeds to a hill in a 12-inch-diameter circle, and later thinned to the two best plants per hill. To get a jump on the season, start seeds indoors one month before planting. Use peat or pressed-manure pots. Plant two pots to a hill. Remove the first flowers if they appear soon after transplanting.

Give the seedlings moisture during the early growth period only. Hoe carefully around the vines until they begin to lengthen. Then mulch with grass, straw, or leaves. By midsummer winter squash will have set all the fruit it will have time to mature. After that, remove all flowers to conserve plant energy for ripening the harvest. Add generous supplies of compost to hills or rows. Squash may be trained on trellises and fences if sling-style support is used for ripening the fruit.

Corn, melons, and pumpkins all make good companions for squash.

Summer squash is ready to harvest in about 55 days; winter squash, in 150. Pick immature summer squash as needed. Don't let them get too large—club-sized fruit is tougher and not as tasty, and the seeds are bigger. Winter squash is left to harden on the vine until the first frost. Cure winter squash for 24 hours outdoors in the sun after cutting it.

Use summer squash fresh and winter squash cured. Neither type freezes well unless it's part of a casserole or soup. Store cured winter squash in a cool, dry place.

## Swiss Chard

Unlike spinach, for which it often substitutes, Swiss chard will grow throughout the summer without bolting. It is perennial in the South, but is usually planted late for a fall crop because leaves dwindle and toughen in very hot weather. Swiss chard, a close relative of beets, is grown for both its large leaves and its broad, ribbed stems.

One mistake gardeners often make with Swiss chard is to plant too much of it. Since the plants keep growing all season and continually replace the harvested outer leaves, they are never "used up." They just keep on producing. A 15-foot row easily meets the needs of a family of four.

Dependable varieties include FORDHOOK GIANT and LUCULLUS (which have green leaves and white stems), and RHUBARB (which has red stalks).

Chard needs an open, sunny, well-drained location. It grows well in the same sort of soil that produces good lettuce and is reasonably fertile.

It is difficult to estimate the yield of Swiss chard because there are different ways of harvesting. One packet of seed sows 15 to 25 feet of row; 1 ounce plants 100 feet. Remember that if you cut individual leaves, chard will continue to produce new ones.

Stored seed is good for four years. Germination rate is 60 percent or better. To hasten germination, soak the seeds for 24 hours before planting.

Sow seeds early in spring, two weeks before the last frost. Plant 1 inch deep, 4 inches apart in rows 18 inches apart, to prevent shading. As the plants mature, thin them to stand 12 inches apart.

Mulch or hoe as weeds appear. Mulch to help a late crop winter over until spring. Swiss chard has a very deep root and survives winter in many areas if it is mulched well enough.

Good companions for chard include beans, members of the cabbage family, and onions.

Swiss chard reaches full size in 50 to 60 days. Cut outer leaves when they are 7 inches high. Keep cutting as needed every three days through summer and fall.

Use Swiss chard fresh, frozen, or canned. The leaves can be cooked like spinach, and the stems can be treated like asparagus. Or stir-fry both parts as you would celery cabbage.

## Tomatoes

This all-time garden favorite requires relatively little care and produces bumper crops. Tomatoes can be grown throughout the United States and in some parts of Canada, wherever there are three to four months of warm weather. Northern gardeners faced with short, cool summers can grow early-maturing varieties bred especially for those conditions, particularly the SUB-ARCTIC type. Tomatoes need to be given protection in areas plagued by hot, dry winds.

Tomatoes come in two different growing patterns. Determinate plants stop growing when they reach a certain size, and their stocky shape requires no pruning or training. In fact, if you prune these plants you cut back on the number of tomatoes they produce. Indeterminate plants keep on growing until dropping temperatures or disease slows them down. They need to be pruned and trained to keep their rampant growth in bounds.

An almost overwhelming assortment of tomatoes is available, of which only a fraction are listed here. Some short-season varieties are COLD SET, FORDHOOK HYBRID, and SUB-ARCTIC. Midseason choices include the classic MARGLOBE and RUTGERS. Large-fruited types include BEEFSTEAK, BIG BOY, BURPEE'S SUPERSTEAK HYBRID VFN, and PONDEROSA. Cherry varieties are SMALL RED CHERRY and SWEET 100. Other varieties with noteworthy characteristics include DOUBLERICH (high in vitamin C), CARO-RED (rich in vitamin A), BURPEE'S LONG KEEPER (long storing), and ROMA and SAN MARZANO (pear-shaped paste varieties).

Choose a sunny, open, well-drained spot with good air circulation for tomatoes. The best soil is slightly acid, organically enriched sand or clay loam. Fruit will ripen faster in sandy soil.

A 100-foot row of tomatoes will produce 4 bushels. One small packet of seed sows a 100-foot row. You can fit 26 unstaked plants in that amount of space.

Tomato seeds can be stored up to four years. Germination rate is 75 percent or more.

Sowing seed directly in the garden is good for early varieties in the South and often produces healthier plants. To give tomatoes time to ripen before frost in the North, start seeds indoors in flats six or seven weeks before the last expected frost, and plan on setting seedlings outdoors in May when frost is past and the ground is warm. Plant seeds 1 inch apart and ¼ inch deep in flats, using three seeds for every plant needed and thinning as necessary. Keep the flats at 70°F after germination, which takes eight to ten days. Transplant the seedlings when the second set of leaves appears, about two weeks after planting.

Transfer the young plants to another flat or to individual pots. Give them more sun and less water than previously. Seven to nine weeks after seeding, they should be 8 to 10 inches tall and have dark green leaves. Leave the plants outdoors in a cold frame over increasing daytime periods for two weeks to harden them off. At night, return the plants indoors or close the lid of the cold frame. If the plants are in flats, cut the soil into cubes a week before transplanting so roots will repair and harden themselves.

To set seedlings in the garden, dig a hole larger than needed for each plant, and fill in the extra space with sifted compost. Set the pot into the hole (if you are using peat pots), and fill in with soil almost to the first set of leaves. Set soil cubes lifted from flats into the prepared holes so that the cubes come in contact with the compost and soil. Fill in the hole and water well.

Staked and pruned plants should be spaced 18 to 24 inches apart. If indeterminate plants are pruned to three stems and trained to a trellis, allow 24 to 30 inches between plants. Plants that are mulched and allowed to sprawl need 3 to 4 feet of room. Small varieties take less space.

Water tomatoes during transplanting and when the weather is extremely dry. Hoe shallowly, or mulch when the soil warms. Sidedress with compost when the vines are in flower, or use compost "tea," a solution of compost dissolved in water, when watering.

Larger vines usually need some support. Use stakes the size of a broom handle, driven 5 to 8 inches into the ground, or use tomato cages. Put stakes in place before transplanting the young plants to the garden. Tie large plants to the stakes loosely and at intervals. Use soft cloth and sling-type ties to support the plants, looping the cloth strips loosely under the leaf nodes. Trim the plants to two main stems, pinching back all additional branches and side suckers. Pinch back the plant top when it reaches the height you want. If you do not stake your plants, prop them upright with piled mulch or straw to keep the ripening fruit off the ground.

Good companions for tomatoes include asparagus, carrots, cucumbers, onions, and peppers.

Tomatoes mature from seed in about 115 days. Pick when they are red and firm (or deep yellow, if you have a yellow variety) but beginning to soften, usually about six days after you notice the first red color. Twist the fruit from the stem, bracing the plant to avoid damage, or cut off the fruit with a sharp knife. Most tomatoes can survive a light frost if well protected by mulch and covers. Ripen full-sized green tomatoes indoors by setting them in a warm (79°F) place. Contrary to common belief, light does not hasten ripening. Unless they are entirely green, the tomatoes will ripen eventually if kept warm.

Serve tomatoes fresh, sliced, or in salads. Stew them, make them into juice, or simmer them into sauce. Green tomatoes can be used for relish or condiments. Tomatoes freeze poorly, but can be canned in different forms.

## Turnips

In the North, turnips are grown as spring or fall crops. In the South they are grown all winter and into early spring. Both the tops and roots are edible.

Turnip greens are all to often overlooked in kitchens, especially outside the Deep South, but they shouldn't be. They provide six times the vitamin C found in the roots. The greens are also rich in vitamins A, B, and E. They can be used raw in salads but most people prefer them cooked.

The rutabaga, or Swedish turnip, is grown in a similar manner but requires longer—90 days—to mature. Rutabagas can withstand more frost than turnips.

Reliable turnip varieties include SEVEN TOP (grown for its leaves), SHOGOIN, PURPLE TOP WHITE GLOBE, JUST RIGHT, DES VERTUS MARTEAU, EXTRA EARLY PURPLE TOP, TOKYO CROSS, and YELLOW GLOBE.

Turnips grow best in a sunny spot with loose, slightly alkaline soil. A good supply of organic matter is particularly desirable for spring crops.

A 100-foot row will give you about 2 bushels of roots. One seed packet sows 50 feet of row.

Turnip seeds store for up to two years. Germination rate is 70 percent or more. Mix the seed with sand to sow thinly.

Sow turnips in spring as soon as the ground can be dug. Prepare the furrows the preceding fall to plant even earlier in spring. For a fall crop, plant two months before you expect the first fall frost. Make spring plantings ¼ inch deep and cover the seeds with sifted compost. Make fall plantings ½ inch deep and cover the seeds with soil. Allow 3 to 4 inches between plants, and 18 inches between rows.

Water turnips in dry weather, and be sure to give them a deep soaking. Mulch lightly in spring and heavily in fall if you plan to store some turnips right in the row during winter.

Fall turnips can take the place in the garden of any crop that matures before the end of July. Peas are good companions for turnips.

Turnip roots mature in 50 to 70 days. The tops are ready to cook in 40 days. Pull and use spring turnips when the roots are 2 to 3 inches in diameter. In the North, dig fall crops before the ground freezes, or mulch heavily. In all but extremely cold locations, mulched turnips can be left in the garden and used as needed.

Turnips are best used fresh, either during the season or from storage. Remove the tops and store unblemished roots in a root cellar, unheated basement, or other cool (32°F), damp, dark place.